Planet In Our Hands:

A Simple Guide to Saving the Earth

Planet In Our Hands: A Simple Guide to Saving the Earth

Maxwell R. Wynter

To my wife Anne ... and Shelley

Table of Contents

Introduction 1

Chapter 1: 5

The Birth of the Universe, Our Solar System, and Earth's Cataclysmic Beginnings

Chapter 2: 11

The Journey of Homo Sapiens

Chapter 3: 17

The Four Horsemen and Two New Ones

Chapter 4: 23

The Architects of War

Chapter 5: 29

The Warming Tide

Chapter 6: 39

The Technology of Today: A Turning Point for Humanity

Chapter 7: 45

Misinformation and the Battle for Truth

Chapter 8: 53

What We Have Learned and The Moment of Reckoning

Chapter 9: 59

The Power of One in a Changing World

Chapter 10: 65

Collective Action: Uniting for Change

Chapter 11: 73

The 100 Year Mission: A Vision for a Reimagined Future

A Note for Those Who Stand at the Beginning 81

Appendix: 83

Reflections on Faith, Science, and Humanity

About Max Wynter 89

Bibliography 91

Introduction

We live in extraordinary times. Our world is changing faster than ever, and with that change comes both remarkable opportunity and unprecedented danger. It's easy to feel overwhelmed. Every day, we're bombarded with information—news articles, social media posts, opinion pieces—each telling us something different about what's happening around us. We're drowning in facts, opinions, and warnings, and yet, as individuals, we feel less and less equipped to do anything about the massive challenges facing humanity.

That's where this book comes in. I wanted to write something different, something that cuts through the noise. This book is not meant to overwhelm you with details or impress you with technical jargon. It's meant to give you a clear and honest picture of where we've come from, where we are, and where we're headed as a species—and more importantly, what you, as an individual, can do about it. This isn't a textbook or an academic tome; it's a guide, a tool to help you navigate the complex problems of our age.

Planet in Our Hands:

There are three main goals here:

1. **Summarize our history**: Understanding where we come from helps us understand where we are. I'll give you the essentials about how humanity evolved and what major events in our history have brought us to this point.

2. **Explore our current situation**: We face many threats—climate change, resource depletion, social inequality, technological disruption, misinformation and more. I'll explain these in clear, accessible language. My goal is not to scare you but to give you a realistic understanding of the issues we face today.

3. **Present solutions**: This is not just a book about problems. I believe in hope, in action, and in the power of individuals to make a difference. I'll walk you through practical, realistic solutions—things we can do to ensure a better future, both as individuals and as a global community.

Now, let's talk about something we all face: **time**. Or rather, the lack of it. In this modern world, finding the time to sit down and read a lengthy book can feel impossible. We're busy, distracted, and constantly pulled in different directions. That's why I've worked hard to make this book concise. You can read it in a single sitting, maybe two. It's under 100 pages for a reason—because I know how hard it can be to get through anything longer, especially when life is pulling you in a hundred different directions. My goal is to give you the maximum amount of information in the shortest time possible, without sacrificing depth or clarity.

Some might say that a book this short can't really cover the complexity of the issues we're facing. To that, I say: **it's a starting point**. Throughout this book, I'll reference other works that dive deeper into the topics we touch on. If something resonates with you or sparks your curiosity, you'll have a roadmap to explore further. I've read every one of these works, from Carl Sagan's **A Pale Blue Dot** to Elizabeth Kolbert's **The Sixth Extinction**. I can tell you that they're worth your time and you'll find a bibliography of the books that I've referred to at the end of the book. But for now, you've picked up this book because you want something you can digest quickly, and that's exactly what I intend to deliver. Each chapter is self-contained and may be read on its own, but all are related.

It's also worth mentioning that this book isn't about religion, but I recognize that science and religion can sometimes feel at odds. If you find that your personal beliefs are challenged by some of the scientific concepts we'll discuss, I encourage you to take a look at the appendix. There I've included some thoughts on how to reconcile the two, or at least how to approach the conversation in a way that respects both science and faith. My goal isn't to persuade you one way or another, but rather to help you see how the two can coexist, even if they don't always agree. If you are an atheist or if you have no difficulty accepting the science presented in this book, then you have no need to read the appendix. Note too that what I've written in the appendix are my own views and that I am not a theologian. I have no professional training in religion.

Now, let's address the elephant in the room: **AI, Artificial Intelligence**, or as Yuval Harari calls it in his book **Nexus**, "Alien Intelligence". Yes, I used AI to help write this book. Some people might feel uncomfortable with that, but I see AI as a tool—like a calculator or a search engine. It's a way to enhance human capabilities, not replace them, at least, not yet. I've reviewed every word in this book, cross-checked the facts and ensured that the ideas presented are grounded in reality. AI helped me organize and clarify my thoughts, but this book is ultimately a reflection of the hundreds of hours I've spent reading, researching, and thinking about these topics. AI isn't a substitute for human effort; it's an amplifier of it. In the same way, I believe that AI can help us tackle some of the world's biggest challenges—but only if we use it wisely. There is a more fulsome discussion about AI in this book.

Speaking of challenges, let's get back to the reason you're reading this. The problems we face today are immense. From the climate crisis, which Bill Gates tackles in **How to Avoid a Climate Disaster**, to the ethical dilemmas posed by genetic engineering, explored in Jennifer Doudna's **A Crack in Creation**, we are living in a time of tremendous

Planet in Our Hands:

risk and opportunity. The choices we make in the next few decades will determine whether we thrive as a species or fall into decline.

It's easy to feel small in the face of these global problems. But I believe in the power of individuals. Katherine Hayhoe, in her book **Saving Us**, makes a compelling case that even small actions can make a big difference. Whether it's changing how you vote, how you consume, or how you communicate with others, there are steps we can all take to move the needle. This book will highlight some of those actions, but the real power lies in you—the reader.

As we move through these pages together, I hope you'll come away with a clearer understanding of our shared history, a sharper awareness of the challenges we face today, and a deeper commitment to the solutions that will shape our future. We are at a critical juncture in human history. The decisions we make today will echo for generations. But you are not powerless. Every small action counts, and I believe that by the time you finish this book, you'll feel empowered to take the first step toward making a difference.

Let's begin.

Chapter 1:

The Birth of the Universe, Our Solar System, and Earth's Cataclysmic Beginnings

The story of everything began roughly 13.8 billion years ago, at a time when there was no space, no time, no matter—just pure energy packed into a single point. And then, in an instant, everything changed. This moment, known as the Big Bang, marked the beginning of the universe as we know it. In a violent explosion, energy expanded outward, cooled, and began forming the basic building blocks of matter: quarks, electrons, and protons. In the vastness of space, stars began to ignite and galaxies swirled into existence.

Among these billions of galaxies was the Milky Way, home to countless stars, but one in particular would come to play a special role: our Sun.

The Formation of Our Solar System

The Sun formed about 4.6 billion years ago, from the gravitational collapse of a massive cloud of gas and dust. As the cloud spun and condensed, it ignited, creating the star that would light our days for billions of years to come. Around it, remnants of that gas and dust coalesced into smaller bodies, forming planets, moons, asteroids, and comets. Earth took shape as one of these planets, emerging from a series of violent collisions in the protoplanetary disk. Over millions of years, Earth grew larger as it absorbed debris from space.

Planet in Our Hands:

Early Earth was a chaotic, molten landscape, constantly bombarded by asteroids and comets. But these collisions weren't just destructive—they were also essential. One of these impacts, occurring about 4.5 billion years ago, likely resulted in the formation of the Moon when a Mars-sized body slammed into the still-forming Earth.

As Earth cooled, its surface solidified, and water, likely delivered by comets and volcanic activity, began to collect, creating the early oceans. Volcanic eruptions spewed gases like carbon dioxide, nitrogen, and water vapor, forming a primitive atmosphere.

Life and Its Fragile Beginnings

For hundreds of millions of years, Earth remained a hostile, lifeless world. But somewhere in the depths of the ocean, about 3.8 billion years ago, simple life forms emerged. Microscopic organisms, likely resembling today's bacteria, developed in an environment driven by volcanic activity and undersea vents. These first forms of life thrived in extreme conditions, and over the next billion years, they evolved and diversified.

Around 2.5 billion years ago, a monumental shift occurred: photosynthetic organisms began to evolve. These tiny microbes, known as cyanobacteria, started producing oxygen as a byproduct of photosynthesis, releasing it into the atmosphere. This was a game-changing development, as it paved the way for more complex life forms to eventually arise.

However, the introduction of oxygen also triggered a massive die-off of many anaerobic organisms that couldn't survive in this new oxygen-rich world. This event, known as the "Great Oxygenation Event", was one of Earth's first mass extinctions, though life quickly adapted and thrived in its wake. While it was a 'mass extinction' it is not generally counted as one of the 'big five' mass extinction events as it is not easy to survey the level or abundance of microscopic organisms or fossil remains from that time. There is enough evidence though to confirm that photosynthesis started, oxygen was produced and a significant number of organisms died out as a result.

The Mass Extinction Events

Earth's history has been shaped by a series of mass extinction events, each radically altering the course of life on the planet. While these extinctions wiped out countless species, they also cleared the way for new forms of life to emerge and dominate.

There have been five major mass extinction events that are well documented:

1. **The Ordovician-Silurian Extinction (around 445 million years ago)**: This was the first major mass extinction, wiping out about 85% of all marine species. It's believed that a period of global cooling, likely triggered by the movement of continents and changes in ocean currents, caused widespread glaciation. Temperatures dropped sharply, freezing much of the planet and drastically lowering sea levels. Species that relied on shallow, warm waters were particularly hard hit.

2. **The Late Devonian Extinction (around 375-360 million years ago)**: This event saw the extinction of about 75% of all species. While the exact cause is debated, it is likely linked to a series of environmental changes, including global cooling, sea level fluctuations, and possibly the impact of large volcanic eruptions. Life in shallow seas was particularly affected, as much of the biodiversity of the time was concentrated there.

3. **The Permian-Triassic Extinction (around 252 million years ago)**: The deadliest mass extinction in Earth's history, the Permian-Triassic event wiped out about 96% of all marine species and 70% of terrestrial species. Known as "The Great Dying," this extinction is thought to have been triggered by massive volcanic eruptions in what is now Siberia. These eruptions released huge amounts of carbon dioxide and sulfur dioxide, leading to global warming, ocean acidification, and widespread anoxia (oxygen depletion in the oceans). Temperatures soared, making Earth nearly uninhabitable for millions of years.

4. **The Triassic-Jurassic Extinction (around 201 million years ago)**: This extinction event is less understood, but it led to the disappearance of about 80% of species. It's believed that volcanic activity and the release of greenhouse gases caused global temperatures to rise, disrupting ecosystems and leading to the collapse of many species. However, this extinction also paved the way for the dominance of dinosaurs, which would rule Earth for the next 135 million years.

Planet in Our Hands:

5. **The Cretaceous-Paleogene Extinction (around 66 million years ago**: Perhaps the most famous mass extinction, this event marked the end of the age of dinosaurs. It was triggered by a massive asteroid impact in what is now the Yucatan Peninsula in Mexico. The impact released an enormous amount of energy, creating a global dust cloud that blocked sunlight, causing temperatures to plummet and photosynthesis to halt. The loss of food chains, combined with volcanic activity and changing sea levels, led to the extinction of about 75% of all species, including the non-avian dinosaurs.

Climate: A Rollercoaster of Heat and Ice

Throughout Earth's history, the planet's climate has swung between extreme heat and cold. These climate shifts have shaped the evolution of life and the fate of species.

One of the most well-known periods of intense cold was the **Cryogenian Period**, which occurred between 720 and 635 million years ago. During this time, Earth experienced one of its most extreme ice ages, sometimes referred to as "Snowball Earth." Temperatures plummeted, and ice sheets extended all the way to the equator. It's thought that much of the planet's surface was frozen over, though life persisted in pockets of open water or beneath the ice.

On the other end of the spectrum were periods of intense heat. For example, during the **Paleocene-Eocene Thermal Maximum (PETM)**, which occurred around 56 million years ago, global temperatures rose by 5 to 8 degrees Celsius. This rapid warming, likely triggered by the release of massive amounts of carbon dioxide and methane, led to significant disruptions in ecosystems. Tropical species migrated toward the poles, while many species in the deep ocean went extinct due to a lack of oxygen.

These temperature extremes had profound impacts on life, driving evolutionary adaptations and shaping the course of species' survival. The swings between ice ages and warm periods—like the **Quaternary Ice Age**, which began around 2.6 million years ago—are a testament to the delicate balance that governs Earth's climate. Human civilizations, for example, have largely thrived during the relatively stable climate of the Holocene Epoch, which began around 11,700 years ago at the end of the last ice age.

The Last Mass Extinction: A Turning Point

The most recent mass extinction, the Cretaceous-Paleogene event, was a pivotal moment in Earth's history. The asteroid impact that triggered this extinction was a cataclysm on a scale humanity can hardly imagine. The blast would have released energy equivalent to billions of atomic bombs, creating shockwaves that rippled across the planet. Fires ignited worldwide, and debris from the impact was ejected into space, falling back to Earth as molten rock. The resulting "nuclear winter" caused temperatures to drop dramatically.

For years, life on Earth struggled to recover from the devastation. With the dinosaurs gone, mammals, which had been relatively small and insignificant, began to rise to prominence. These survivors took advantage of the empty ecological niches left behind, leading to the evolution of a wide variety of species, including the primates that would eventually give rise to humans.

In the grand sweep of history, mass extinctions have served as both destroyers and creators. They wipe the slate clean, but in doing so, they provide opportunities for new life to flourish.

And so, Earth continued its journey, shaped by forces beyond its control, carrying the seeds of life that would one day become us.

Planet in Our Hands:

Chapter 2:

The Journey of Homo Sapiens

Sixty-six million years ago, the world as we know it changed. A fiery asteroid smashed into Earth near the Yucatán Peninsula, marking the end of the Cretaceous period and wiping out the dinosaurs. What followed was the Cretaceous-Paleogene extinction event, the fifth mass extinction in Earth's history. However, amid the chaos, life found a way. Small, furry mammals, no larger than today's house cats, scurried underground, surviving on insects and plants. It was from these survivors that all modern mammals, including us, Homo sapiens, would eventually evolve.

In the shadow of the dinosaurs' demise, the Age of Mammals began. The creatures that survived the catastrophe weren't the large, dominant species but the adaptable, smaller ones that could live underground, avoid danger, and survive off a variety of food sources. Over millions of years, these early mammals evolved, diversifying into various species that would fill the ecological void left by the dinosaurs.

Among these species, one lineage would give rise to primates—creatures who developed larger brains and more dexterous hands, a trait that would become essential for their survival. As Earth's climates shifted and forests gave way to open savannas, some of these primates, particularly in Africa, stood upright. Walking on two legs freed their hands, allowing them to use tools and develop language. And so, the first humans—hominins—were born.

The Rise of Homo Sapiens

Fast forward to about 300,000 years ago in Africa, where Homo sapiens, the species we belong to, made their grand entrance. But we weren't alone. Earth was home to other human species, like the Neanderthals in Europe and the Denisovans in Asia. These species were similar to us in many ways, but Homo sapiens had something special. Our brains allowed for complex language, creativity, and problem-solving skills.

Homo sapiens were curious and restless. While other human species remained in their ecological niches, Homo sapiens pushed the boundaries. Around 60,000 years ago, our ancestors left Africa, driven by a combination of necessity and curiosity. What followed was the most remarkable migration in the history of life on Earth.

With each step, we spread out, first to the Middle East, then Europe, Asia, and eventually as far as Australia and the Americas. This journey was not without consequence. As Homo sapiens arrived in new lands, ecosystems were irreversibly altered. Large animals, like the woolly mammoths in North America and the giant kangaroos in Australia, suddenly faced a new predator: humans armed with intelligence and tools.

The migration of Homo sapiens triggered one of the earliest human-driven extinctions. As we spread, we hunted many species to extinction, contributing to the sixth mass extinction event that continues to unfold today. But our expansion wasn't only about survival; it was about dominance. Homo sapiens didn't just adapt to their environments—they transformed them, often with devastating consequences for other species.

The Stone Age: A New Dawn

In the vast plains and forests of prehistoric Earth, Homo sapiens learned to adapt to their environments in increasingly sophisticated ways. The first tools were simple stone implements used to butcher meat, but over time, our ancestors became more inventive. The Stone

Age, which began around 2.5 million years ago, saw the gradual refinement of these tools, giving humans an advantage over other predators.

These tools were more than just practical. They marked the beginning of human culture. Early Homo sapiens began creating art, as evidenced by cave paintings in places like Lascaux, France, and the Blombos Cave in South Africa. The ability to create symbols and art suggested something profound—humans were developing the ability to think abstractly.

With the invention of fire, Homo sapiens gained even more control over their environment. Fire allowed us to cook food, providing more energy for our growing brains. It also enabled humans to venture into colder climates, giving them the ability to survive in a wide variety of habitats. It wasn't just about surviving anymore—it was about thriving.

Out of the Stone Age: Iron, Bronze, and the Age of Cities

As human populations grew and resources became more abundant, our ancestors transitioned out of the Stone Age and into the ages of metal. Around 3300 BCE, the Bronze Age began. It was during this time that humans discovered how to combine copper with tin to make bronze, a stronger and more durable material than stone. Bronze tools and weapons allowed civilizations to grow, leading to the formation of the first cities.

The Iron Age followed around 1200 BCE, and with it, even more technological advancements. Iron tools revolutionized agriculture, warfare, and transportation. Societies could now build large empires, as we saw with the Egyptians, the Romans, and the Persians. It was also during this time that writing systems were developed, allowing humans to record history, laws, and stories.

The rise of these ancient civilizations also gave birth to the world's first organized religions. As humans grappled with questions about their place in the universe, they turned to gods and myths for answers. Religion provided not only a sense of order but also a way to explain the natural world. In ancient Mesopotamia, Egypt, and Greece, religious

Planet in Our Hands:

beliefs were intertwined with political power, shaping the course of history.

The Role of Religion in Human Evolution

Religion, more than any other institution, shaped the intellectual growth of Homo sapiens. As Yuval Harari notes in **Sapiens**, it was religion that united large groups of people under shared beliefs, making it possible to organize and build large civilizations. Religion helped humans cooperate on a scale never seen before in the natural world.

The early belief systems were deeply tied to nature. Many early religions, like those in ancient Mesopotamia and Egypt, saw the forces of nature—storms, floods, and droughts—as controlled by gods. Humans, dependent on the land for survival, sought to appease these gods through rituals and sacrifices.

As civilization advanced, so did religion. Monotheism, the belief in a single god, began to take hold, especially with the rise of Judaism, Christianity, and Islam. These religions not only offered spiritual guidance but also brought with them complex systems of morality, justice, and law. They provided a framework for how humans should live, and in doing so, influenced everything from politics to art to education.

Ecological Changes and the Anthropocene

By the time humans had mastered agriculture and metallurgy, they had become the dominant species on the planet. But with great power comes great responsibility—or in this case, great destruction. The spread of Homo sapiens reshaped ecosystems across the globe. As hunter-gatherers, our impact was relatively small, but the advent of agriculture around 10,000 years ago changed everything.

The development of farming allowed human populations to grow rapidly, but it came at a cost. Forests were cleared, rivers were dammed, and animals were domesticated to serve our needs. In the process, many ecosystems were destroyed. This marked the beginning of the Anthropocene, the age in which humans became the dominant force shaping the planet.

The impact of Homo sapiens on the planet is profound. As Elizabeth Kolbert explains in **The Sixth Extinction**, our species has caused more extinctions than any other in Earth's history. From the dodo to the passenger pigeon, countless species have been wiped out as a result of human activities. And while we have made incredible technological advancements, the cost has been the degradation of Earth's ecosystems.

The Intellectual Leap: Science and Knowledge

Through the Stone Age, Bronze Age, and Iron Age, humans developed increasingly complex tools and societies. But it wasn't until much later that Homo sapiens began to unlock the secrets of the universe. It was during the Renaissance in the 15th century that humans made a great intellectual leap, sparked by the rediscovery of ancient Greek and Roman texts. Thinkers like Galileo, Newton, and Kepler questioned the old ways of thinking, and in doing so, laid the foundations for modern science.

Stephen Hawking, in **A Brief History of Time** and **The Theory of Everything**, explores the universe's mysteries in a way that builds on the work of these early scientists. It's an astounding journey that began with simple tools and fire, leading us to the discovery of black holes, quantum mechanics, and the very nature of time itself.

As we moved out of the Iron Age and into the modern world, religion's grip on intellectual thought loosened, replaced by a new faith: science. And yet, the two remain intertwined. While science provides us with explanations for how the world works, religion still offers many answers for why we are here.

The Future of Homo Sapiens

Looking back at our journey from tiny mammals scurrying in the shadows of dinosaurs to the dominant species on Earth, it's clear that Homo sapiens have always been defined by our adaptability. We are a species of inventors, explorers, and thinkers. But as we look to the future, we must grapple with the consequences of our actions.

The sixth mass extinction, climate change, and the degradation of natural resources are all challenges we created, and now they threaten

Planet in Our Hands:

our survival. But if history has taught us anything, it's that Homo sapiens have a remarkable ability to adapt and overcome.

The question is, will we? Our survival will depend on our ability to learn from the past and chart a course toward a sustainable future.

Chapter 3:

The Four Horsemen and Two New Ones

In ancient times, the world shuddered at the mention of the Four Horsemen: War, Famine, Disease, and Death. These riders, symbols of catastrophe, charged through civilizations, leaving scars that took generations to heal. They rode across the earth, bringing chaos, despair, and suffering. Yet, as the world evolved, so did humanity's response.

As the 20th and 21st centuries unfolded, mankind faced these ancient forces with newfound strength. The Horsemen, once unstoppable, met resistance from the most intelligent, creative, and determined species the planet had ever seen. Humanity pushed back, armed with knowledge, technology, and a resolve to build a better world. Though the battle came at a cost, humanity's progress was undeniable.

However, as progress marched on, two new riders emerged: Climate Change and Artificial Intelligence. Unlike their predecessors, these modern Horsemen posed threats not just to human bodies but to the entire planet and the essence of human society itself. Misinformation, riding pillion behind Artificial Intelligence, added another layer of complexity, undermining truth and trust. The challenge grew: could humanity conquer the physical dangers of the past while also facing the emerging, unseen threats of the future?

Planet in Our Hands:

War: An Unending Struggle

War, the most brutal of all Horsemen, has long been humanity's tormentor. The 20th century saw two devastating world wars, and even in the 21st century, conflicts rage on. The invasion of Ukraine, for example, shows how the ambitions of powerful leaders can drag nations into battle, shattering lives and reshaping borders.

Despite this, progress has been made. International organizations like the United Nations emerged, striving to replace violence with diplomacy. Peacekeeping missions, treaties, and alliances have managed to stave off the full-scale conflicts that once seemed inevitable. The Nuclear Non-Proliferation Treaty is one such example, keeping humanity from crossing lines that could have led to catastrophic consequences.

Yet, the threat of war remains a constant presence. The actions of a few powerful individuals or nations can still bring destruction, reminding us that while much has been achieved, the battle to defeat war is ongoing. The goal is clear: find new ways to promote peace, dialogue, and cooperation so that future generations can live free from the shadow of war.

Famine: Feeding a Growing World

Famine has haunted humanity for centuries, bringing suffering and despair to millions. In the past, crop failures, droughts, and resource scarcity led to widespread hunger. But the advancements of the 20th and 21st centuries sparked a revolution in food production. The Green Revolution which started in the 1940's, for instance, transformed agriculture with improved crop varieties, fertilizers, and irrigation techniques, lifting billions out of hunger and transforming entire societies.

Genetic engineering, especially through the new CRISPR technology, offers new possibilities. It enables the development of crops that can withstand harsh climates and resist pests, reducing the need for harmful chemicals and making it possible to feed a growing population despite the challenges of a changing environment. The dream of a world without hunger is no longer a distant hope but a reachable reality.

However, famine is not just a technological issue; it's also a matter of power. Even today, conflicts and political struggles disrupt food supplies, turning fields into battlegrounds and making it hard to deliver aid to those in need. In some cases, food becomes a weapon, used by powerful leaders to control or punish populations. While humanity has the capability to produce enough food for everyone, the actions of a few continue to threaten the many.

Disease: The Pursuit of Health

The third Horseman, Disease, has left a devastating mark on human history. Two from our not-so-distant past were the Plague of Justinian which killed an estimated 15 to 100 million people in the 6th century and the Black Death, killing 25 to 50 million in the 14th century. In the early 20th century, the Spanish flu also killed millions. Nearly a century later, COVID-19 swept across the globe, highlighting humanity's vulnerability but also its resilience. While the Spanish flu claimed over 50 million lives, the death toll of COVID-19, although tragic, was significantly lower due to advancements in medical technology and global cooperation.

Vaccines, antibiotics, and modern healthcare systems have transformed humanity's battle against disease. What once seemed like acts of fate can now be managed, controlled, and sometimes even eradicated. Polio, once a terrifying threat, is now nearly extinct, thanks to worldwide vaccination efforts. The same scientific breakthroughs that mapped the human genome are now being used to develop personalized medicine, offering treatments that were unimaginable in the past.

CRISPR technology also brings hope to the field of medicine. It offers the possibility of curing genetic disorders and eliminating diseases like HIV. This power to edit genes could reshape human health, allowing the eradication of conditions that have plagued humanity for centuries.

However, new diseases will continue to emerge, and old ones may return. The fight against the Horseman of Disease is a race without a

Planet in Our Hands:

finish line, requiring constant vigilance, innovation, and international collaboration.

Death: Defying the Ultimate Boundary

Death, the final Horseman, remains humanity's inevitable companion. But instead of accepting it as a certainty, humanity has fought back. Advances in medicine, public health, and biotechnology have extended life expectancy across the globe. In the 21st century, researchers like David Sinclair are exploring ways to slow aging, turning longevity from a dream into an achievable goal. By understanding the biological mechanisms of aging and developing treatments that target them, the idea of living longer and healthier lives is becoming a reality.

CRISPR again plays a central role, with the potential to repair or eliminate age-related diseases. This isn't just about adding years to life; it's about ensuring those extra years are filled with health and vitality. The idea that death could be delayed—perhaps indefinitely—sparks both excitement and caution.

Ethical questions arise. Who will have access to these breakthroughs? Will they be available only to the wealthy, or will they be shared with all? The quest to extend life brings with it the responsibility to ensure fairness and equity.

I have taken the liberty of creating two new horsemen, Climate Change and Artificial intelligence.

Climate Change: The Planet's Fury

Climate Change, the first of the new Horsemen, poses a threat unlike any other. It affects not just one region or nation but the entire planet. Rising temperatures, extreme weather events, and melting ice caps are already reshaping the world, threatening coastal communities and pushing ecosystems to the brink. This Horseman is not a future danger; it is here, now.

Efforts to combat Climate Change are underway. Green energy sources like wind, solar, nuclear, and hydropower offer alternatives to fossil fuels. Initiatives to protect and restore forests and wetlands aim to balance the carbon cycle. Global agreements, such as the Paris Accord, set targets for reducing emissions. But progress is slow, and powerful interests—oil companies, political leaders, and corporations—sometimes stand in the way.

Climate Change challenges humanity's ability to act collectively. It's a test of whether people can unite, not just within their own nations but across borders, to protect the planet they share. The battle is not just about reducing emissions but also adapting to the changes that are already occurring and finding ways to build a sustainable future for all.

Artificial Intelligence: A Double-Edged Sword

The final Horseman, Artificial Intelligence (AI), rides alongside Climate Change, wielding both promise and peril. AI has the potential to revolutionize everything from healthcare to agriculture, offering solutions that could help humanity defeat the other Horsemen. It can analyze massive amounts of data, identify patterns, and create strategies for disease prevention, food production, and even climate mitigation.

However, AI also brings risks. Autonomous weapons, biased algorithms, and the potential for mass surveillance loom large. Moreover, misinformation—often driven by AI algorithms—spreads quickly, undermining trust in institutions and facts. Social media, powered by AI, has become a battlefield where truth and lies clash, sometimes with deadly consequences. The COVID-19 pandemic highlighted this danger, as false information about vaccines spread faster than the virus itself.

The challenge with AI is to harness its potential for good while minimizing its risks. Governments and tech companies must work together to create ethical guidelines, regulate AI development, and prevent its misuse. If left unchecked, this powerful technology could become a tool for control, manipulation, and division.

Looking Ahead

Humanity's fight against the Horsemen continues. While we have made progress in overcoming War, Famine, Disease, and Death, the arrival of Climate Change and Artificial Intelligence adds new challenges that require global action, innovation, and unity. Mankind's achievements in

Planet in Our Hands:

the 20th and 21st centuries show what is possible when people come together, but the battle is far from over.

The next chapters will explore in depth the threats posed by War, Climate Change, Artificial Intelligence and AI's pillion rider Misinformation, delving into their complexities and the steps needed to address them. The story of humanity is one of resilience and progress, but the path forward requires courage, cooperation, and a commitment to the truth. The question remains: will we rise to the challenge, or will we allow the new Horsemen to run free?

Chapter 4:

The Architects of War

Wars have always shaped the trajectory of human history. For thousands of years, individuals seeking power, resources, or revenge have plunged entire nations into conflict. Each war leaves scars on the land and on the souls of those who survive it, yet the motivations behind these wars are often strikingly similar. As we look at the wars of the past 125 years, we find a disturbing pattern: personal ambition, greed, and a desire for dominance have repeatedly driven leaders to start wars, while the human needs they manipulated—fear, patriotism, and survival—paved the way for violence on a massive scale.

The First World War: National Pride and Ambition

In 1914, the world was thrown into chaos as World War I broke out— a conflict that was supposed to be "the war to end all wars." The spark that ignited this inferno was the assassination of Archduke Franz Ferdinand of Austria, but the fire had been smoldering for years. The true causes of the war lay in the tangled web of alliances, militarism, and nationalism that dominated Europe at the time.

Kaiser Wilhelm II of Germany played a significant role in escalating tensions. He was deeply insecure about Germany's place on the world stage and believed that war could secure its dominance in Europe. Wilhelm's ambitions were fueled by a need to prove that his nation was superior to its rivals, particularly France and Britain. As Yuval Noah

Planet in Our Hands:

Harari explains in **Sapiens**, humans have always felt a need to create and defend in-group identities, and Wilhelm used this instinct to rally support. He stoked fears of German vulnerability, painting the war as necessary for Germany's survival. Yet in the end, the war decimated Europe, and Wilhelm's Germany was left defeated, humiliated, and eventually impoverished.

The Treaty of Versailles that followed was meant to ensure lasting peace, but it did just the opposite. The harsh terms imposed on Germany created resentment and paved the way for another war. In the quest for pride and power, Wilhelm accomplished little but set the stage for future conflicts.

World War II: Hitler's War of Ideology and Revenge

Adolf Hitler was one of the most infamous figures in modern history, responsible for unleashing the horrors of World War II. His rise to power in Germany was fueled by the anger and economic hardship that followed World War I. Hitler saw himself as a savior, promising to restore German pride and avenge the humiliations imposed by the Treaty of Versailles. His ideology was based on a toxic combination of racism, nationalism, and expansionism.

Hitler used fear as a primary tool to manipulate the German people. The economic collapse of the 1930s had left millions of Germans desperate for change, and Hitler provided them with a scapegoat: the Jewish people. By convincing the population that their suffering was due to an "enemy within," Hitler was able to unite the country behind his genocidal plans.

The war he started in 1939 was driven by a desire for "Lebensraum," or living space, for the German people. Hitler envisioned a vast empire stretching across Europe, but his ambitions were ultimately unsustainable. Millions of people, soldiers and civilians alike, perished in the conflict. Germany was left in ruins, divided, and occupied by foreign powers. In his quest for dominance, Hitler achieved only death and destruction, both for his enemies and his own people.

The Cold War: Ideological Rivalries and Proxy Wars

The aftermath of World War II led to a new kind of conflict—one fought not with direct military engagement, but through influence, espionage, and proxy wars. The Cold War between the United States and the Soviet Union was driven by two conflicting ideologies: capitalism and communism. Both nations saw themselves as champions of their respective ways of life, and both feared the other's influence on the world.

Leaders like Joseph Stalin and later Nikita Khrushchev in the Soviet Union, and various U.S. presidents like Harry Truman, Dwight Eisenhower, and John F. Kennedy, viewed the Cold War as a battle for global supremacy. While there were certainly geopolitical and economic reasons behind the conflict, much of it was about maintaining power and influence over the world stage.

The Cuban Missile Crisis of 1962 exemplified how close the world came to total destruction. Khrushchev's decision to place nuclear missiles in Cuba was a direct challenge to American dominance in the Western Hemisphere. The fear of nuclear annihilation hung over the heads of citizens across the globe. While the standoff was resolved without direct conflict, it highlighted the dangers of two superpowers locked in an ideological struggle.

Throughout the Cold War, proxy wars in places like Vietnam, Korea, and Afghanistan were fought in the name of ideological supremacy. The leaders of the superpowers manipulated local conflicts to serve their own interests, leading to decades of devastation and instability in those regions. While the Cold War itself ended with the dissolution of the Soviet Union in 1991, the scars of those proxy wars remain today.

The War on Terror: Fear and Control

On September 11, 2001, Al-Qaeda, a terrorist organization, attacked the United States of America by destroying the twin towers of the World Trade Centre in New York City and a part of the Pentagon building, headquarters of the United States Armed forces. In the wake of these attacks, the United States launched a global "War on Terror." The

Planet in Our Hands:

primary target was Al-Qaeda, the terrorist organization behind the attacks, but the scope of the war quickly expanded. President George W. Bush framed the conflict as a battle between good and evil, using the rhetoric of fear to gain public support for military interventions in Afghanistan and Iraq.

While the initial invasion of Afghanistan was aimed at dismantling Al-Qaeda and the Taliban regime that harbored them, the war in Iraq was driven by different motivations. The Bush administration argued that Iraq's dictator, Saddam Hussein, possessed weapons of mass destruction and posed a threat to global security. Yet, no such weapons were ever found, leading many to conclude that the war was more about securing oil resources and asserting American dominance in the Middle East.

The War on Terror has had lasting consequences. The rise of ISIS, another terrorist organization, ongoing instability in Iraq and Afghanistan, and the displacement of millions of people are just a few of the war's outcomes. While the U.S. did eliminate some terrorist leaders, including Osama bin Laden, head of Al-Qaeda, the war itself has not eradicated terrorism. Instead, it has fueled anti-American sentiment and contributed to the radicalization of new generations.

The Russian Invasion of Ukraine: Power and Expansionism

In 2022, Russia's President Vladimir Putin launched a full-scale invasion of Ukraine, a move that shocked the world. Putin's motivations for the invasion were complex but centered around his desire to reassert Russian influence over former Soviet territories. Putin framed the conflict as a defensive measure, arguing that NATO's expansion threatened Russia's security. However, many experts view the invasion as a clear attempt to restore Russia's dominance in Eastern Europe and to undermine Ukraine's efforts to join Western institutions like the European Union.

Putin's actions have had devastating consequences for both Ukraine and Russia. Thousands of civilians have died, cities have been reduced to rubble, and millions of Ukrainians have been displaced. Meanwhile, Russia has faced economic sanctions, international condemnation, and isolation from the global community. The invasion has also triggered a renewed sense of unity in NATO, The North Atlantic Treaty Organization, as countries rally to support Ukraine's defense.

Putin, like many leaders before him, relied on a sense of national pride and fear to justify his actions. He portrayed Ukraine as a threat to Russian identity, tapping into historical grievances and nationalistic fervor. But the war has only deepened Russia's isolation and weakened its economy. It remains to be seen what lasting impact this conflict will have on the global order, but the human cost is already staggering.

The Role of Human Failings in War

What ties all these conflicts together is the role of individual leaders in shaping the course of history. Whether driven by a thirst for power, a desire for revenge, or a need to control resources, these leaders have manipulated their populations to achieve their goals. In each case, they exploited basic human needs—security, identity, and survival—to rally support for their wars. But these wars rarely led to long-term benefits for the people who fought them.

The pursuit of power often blinds leaders to the suffering they cause. As Elizabeth Kolbert explains in **The Sixth Extinction**, human beings have an uncanny ability to ignore the long-term consequences of their actions, especially when motivated by short-term gains. The same can be said for war. Leaders often focus on immediate victories, without considering the lasting harm to their nations and the world.

Could These Wars Have Been Avoided?

In many cases, war seems inevitable in hindsight, but history suggests that diplomacy, compromise, and cooperation could have prevented some of the bloodshed. The failure of the Treaty of Versailles to create a stable post-World War I Europe was a major factor in the rise of Hitler, and more effective diplomacy could have prevented the Second World War. Similarly, the Cold War could have escalated into nuclear annihilation, but cooler heads prevailed during key moments.

The same is true of more recent conflicts. The War on Terror might have taken a different course had world leaders focused more on

Planet in Our Hands:

addressing the underlying causes of terrorism—poverty, political instability, and religious extremism—instead of relying on military force. The Russian invasion of Ukraine might have been avoided through more effective international engagement and a deeper understanding of Russia's security concerns, though it's unclear if anything could have prevented Putin's ambitions entirely. Thinking beyond Russia, is there any logical reason for China to continue its own desire to expand into Taiwan? Is national pride and hegemony enough reason for continuing down a road that can only lead to immense suffering, death and disruption of the global economy. Similar discussions can be had about many disputes, for example Venezuela and Guyana, India and Pakistan and perhaps the biggest powder keg of all, the middle east.

Ultimately, war is often the result of human failings—our greed, our fear, and our inability to see beyond our immediate needs. But as Mustafa Suleyman argues in **The Coming Wave**, the future doesn't have to be shaped by conflict. New technologies, better communication, and a focus on shared global challenges like climate change, discussed in the next chapter, could help us avoid the mistakes of the past. We can learn from history, or we can repeat it. The choice is ours.

Chapter 5:

The Warming Tide

The Earth has always known change. For billions of years, it has cycled through ice ages and warm periods, its surface shifting and evolving. But never has the world seen a shift like this—one not born from natural cycles but from the hands of its own inhabitants. It's as if humanity, in its relentless pursuit of progress, forgot the delicate balance that kept the world in harmony.

To understand where we stand now, we must go back in time—to when the Earth last faced catastrophic changes of this magnitude.

The Origins of Change

Sixty-six million years ago, the planet shook under the impact of an asteroid, plunging the Earth into darkness and wiping out the dinosaurs. It was a cataclysmic event, causing fires that burned for weeks and clouds of ash that blocked out the sun for years. The Earth's temperature plummeted, and life as it was known came to a crashing halt.

Fast forward to the present day, and the threat isn't a giant rock from space—it's us. Elizabeth Kolbert, in **The Sixth Extinction,** warned of a new wave of species disappearing, one caused by a force unlike any before: human activity. We have transformed the landscape with our cities, poisoned the air with our factories, and altered the chemistry of

Planet in Our Hands:

the atmosphere. In just a few centuries, the human race has become a geological force, capable of shaping the destiny of the entire planet.

The Industrial Revolution was a turning point. When factories first roared to life, churning out smoke, it was a symbol of progress, a promise of a better future. But that promise came with a price. Coal, oil, and gas became the lifeblood of modern civilization, powering everything from factories to homes. As these fossil fuels burned, they released carbon dioxide into the atmosphere, trapping heat like a blanket. And slowly, the Earth began to warm.

The trouble is it didn't stay slow for long.

The Alarming Pace

Throughout the Earth's history, warming events have happened but never at the speed we see today. During the Paleocene-Eocene Thermal Maximum (PETM), around 56 million years ago, temperatures rose by 5 to 8 degrees Celsius over several thousand years. Life struggled but adapted. But today, the rate of temperature increase is unprecedented, warming by roughly 1.2 degrees Celsius in just over a century. The Earth's systems, from its oceans to its weather patterns, are struggling to keep up.

To see this transformation in action, one need only look at Isle de Jean Charles, a small island off the coast of Louisiana. For centuries, it was home to Native American communities—people who lived in harmony with the land and waters around them. But today, the island is disappearing. Rising sea levels, intensified storms, and coastal erosion have swallowed 98% of the land since 1955. What was once a vibrant community is now a narrow strip of land, surrounded by brackish water and skeletal remains of trees.

The story of Isle de Jean Charles isn't just about one community. It's a warning. The U.S. Army Corps of Engineers has declared that the island is no longer habitable, and federal funds have been allocated to relocate its residents. Known as America's "first climate refugees," the people of Isle de Jean Charles are being forced to abandon their homes, not because of conflict or economic hardship, but because the land itself is vanishing beneath their feet.

This tragedy, unfolding in one of the world's wealthiest nations, highlights a painful reality: climate change knows no borders. It will affect rich and poor alike, but marginalized communities, like those on Isle de Jean Charles, are often the first to suffer. The island's fate is a microcosm of what coastal regions around the world could face if global warming continues at its current pace.

The Thwaites Glacier: A Tipping Point

As we consider the alarming pace of climate change, one of the most significant threats lies at the heart of Antarctica: the Thwaites Glacier. Often referred to as the "Doomsday Glacier," Thwaites is a massive ice formation roughly the size of Florida. Its stability is crucial to the planet, but it is melting at an accelerating rate. If Thwaites were to collapse entirely and fall into the ocean, it could trigger a chain reaction with catastrophic consequences.

Thwaites acts as a critical buttress for other glaciers in West Antarctica. If it collapses, it could destabilize the surrounding ice, leading to the release of enough ice into the ocean to raise global sea levels by as much as 10 feet (about 3 meters). This is not a distant threat; researchers estimate that significant ice loss could happen within the next few decades if warming continues unchecked.

The impact of such a rise would be devastating. Coastal cities worldwide, from Miami to Mumbai, would face unprecedented flooding, displacing millions of people and causing trillions of dollars in damage. Entire ecosystems could be submerged, and agricultural lands crucial for global food supplies could be permanently lost. The collapse of Thwaites Glacier is a stark reminder that climate change is not just about rising temperatures; it's about the very foundations of our world shifting beneath us.

Jeff Goodell, in **The Heat Will Kill You First**, highlights the urgency of addressing these tipping points before they become irreversible. The Thwaites Glacier serves as a warning—one that, if ignored, could transform the planet in ways humanity may not be able to reverse. But the Thwaites Glacier is just the tip of the Climate Change iceberg.

Planet in Our Hands:

Climate Change: The Fifth Horseman

In the story of humanity's greatest threats, climate change has emerged as the fifth horseman. This horseman doesn't ride alone—it is accompanied by war, famine, and disease, each exacerbated by its presence. Rising temperatures and shifting climates destabilize nations, creating conflict over resources like water and fertile land. Droughts and floods threaten food supplies, leading to famine. And as weather patterns become more unpredictable, diseases spread to new regions, claiming more lives. Climate change isn't just an environmental issue; it's a force that intensifies every other crisis, making it one of the most formidable challenges humanity has ever faced.

The Science Behind the Crisis

Scientists have warned of this moment for decades. In the 1980s, climatologist James Hansen testified before the U.S. Congress about the dangers of unchecked fossil fuel emissions. His models showed a future where heatwaves, fires, and rising seas would become the norm. The data was there, but the will to change was not.

Why? Part of the answer lies in the structure of modern society. As Yuval Noah Harari discusses in **Homo Deus**, humanity has mastered the art of control, manipulating nature to suit our needs. This control has given us incredible power, but it also blinded us to the consequences. The convenience of fossil fuels, the lure of constant economic growth, and the belief that technology could solve any problem led us down this path.

Yet the reality of climate change has proven to be more complex than anticipated. The increase in carbon dioxide doesn't just warm the air; it acidifies the oceans, destroying coral reefs that sustain marine life. As temperatures rise, permafrost in the Arctic thaws, releasing methane—a greenhouse gas far more potent than carbon dioxide—into the atmosphere, accelerating the warming process. This is the feedback loop scientists have long feared: one event triggering another, amplifying the problem beyond control.

Jennifer Doudna's work on CRISPR, as outlined in **A Crack in Creation**, shows how technology can be a double-edged sword. While her gene-editing breakthrough holds promise for curing diseases, the same ingenuity that allows us to manipulate DNA has also enabled the exploitation of natural resources on an unprecedented scale. The question remains: can humanity use its technological power to reverse the damage, or will it only deepen the crisis?

The Dilemma of Developing Nations

As the world grapples with this crisis, a new challenge emerges: the divide between rich and poor nations. Wealthy countries have the resources to invest in renewable energy and adapt to a changing climate. But what about countries that don't? Developing nations often rely on fossil fuels for economic growth, and cleaner alternatives remain out of reach financially.

Guyana, for example, is a small nation on the northern coast of South America. For decades, it struggled with poverty and underdevelopment. But recently, the discovery of vast offshore oil reserves has offered a path to prosperity. The country, one of the poorest in the region, now has the potential to lift its people out of poverty and build a stronger economy. But there's a catch: exploiting these resources means increasing carbon emissions, contributing to the very problem the world is trying to solve. Guyana's problem is even more interesting. At high tide many communities on Guyana's coast are below sea level. To protect them a Sea Wall, approximately 450 kilometers (280 miles) long was built and completed in 1892. I've stood on this wall, which has been breached by the ocean on several occasions. This sea wall protects Guyana's capital Georgetown which will likely be flooded if the Thwaites Glacier slips into the ocean.

This dilemma is not unique to Guyana. Across Africa, Asia, and South America, countries rich in natural resources face similar choices. The pressure to improve living standards and build infrastructure often outweighs the push for cleaner energy, especially when renewable options are costly, and infrastructure is lacking. The international community urges these nations to "go green," but without substantial support, it's a nearly impossible task.

Planet in Our Hands:

Kim Robinson's **The Ministry of the Future,** a work of fiction, explores this tension. He imagines a world where developing nations demand compensation from wealthy countries for the carbon they have historically emitted. It's a fair argument: if the West built its wealth on fossil fuels, why shouldn't poorer countries have the same opportunity? The solution isn't simple, and it requires unprecedented levels of cooperation, funding, and technology sharing to make clean energy viable for all.

A Path Forward

David Sinclair's work in **Lifespan** reminds us of the incredible strides humanity has made in other areas, like healthcare and genetics. The same ingenuity that has extended human life expectancy can be applied to extend the life of our planet. If we harness this power with the urgency and determination required, the climate crisis might be one of humanity's greatest triumphs, not its downfall.

Countries like Guyana can become models for a new kind of growth—one that balances economic development with environmental stewardship. By investing in sustainable practices and innovative technologies, they could show the world that progress doesn't have to come at the expense of the planet. The world's wealthiest nations must stand by their commitments and support these efforts, ensuring that the resources are available for those who need them most.

The story of global warming is not yet finished. We stand at a crossroads, faced with a choice. One path leads to further destruction—a continuation of the same mistakes. The other offers a chance for redemption, where humanity rises to the challenge, applying its knowledge, technology, and unity to heal the planet.

Hope and Solutions

Despite the grim outlook, there is still hope. Scientists, activists, and innovators are working tirelessly to find solutions. Katharine Hayhoe, in **Saving Us**, emphasizes the power of communication and community action. She argues that by building bridges and finding common ground, people can mobilize to demand change, pushing governments and corporations to act.

Technological innovation also offers a lifeline. Mustafa Suleyman's **The Coming Wave** highlights the potential of artificial intelligence in tackling climate change. AI can optimize energy usage, design smarter cities, and even predict weather patterns, allowing for more effective disaster response. Renewable energy sources like solar and wind are becoming more efficient, offering a path away from fossil fuels. If scaled up, these technologies could dramatically cut emissions. However, making these solutions accessible to all, especially developing nations like Guyana, remains a challenge.

Bill Gates proposes investing in breakthrough innovations like carbon capture technology, which can remove carbon dioxide directly from the atmosphere. This technology, while still in its infancy, could be a game-changer if it becomes economically viable and accessible to nations like Guyana. Moreover, there is a growing recognition that developed nations, which have historically been the largest polluters, bear a responsibility to support developing countries. This includes not only financial assistance but also transferring technology and expertise. Collaboration could lead to innovations that are cost-effective and suitable for the specific needs of developing regions—like solar energy projects that power remote communities without the need for an extensive grid.

The Thwaites Glacier and Global Efforts

The Thwaites Glacier serves as a critical tipping point that underscores the urgency for collective action. International efforts to prevent its collapse and mitigate sea level rise could set the tone for broader cooperation. Scientists are exploring ways to slow its melting by reinforcing the glacier with technologies like underwater barriers to block warm ocean currents. However, such efforts require massive funding and international coordination.

Countries must work together through platforms like the United Nations Climate Change Conferences (COP) to create binding agreements and secure commitments from the world's biggest polluters. At the same time, investments must be made to protect vulnerable communities, such as those living near Isle de Jean Charles or

Planet in Our Hands:

developing countries like Guyana, Bangladesh and Chad from the worst effects of climate change.

A Global Movement

In the face of these challenges, a global movement is growing. Activists like Greta Thunberg have rallied millions of young people worldwide, demanding action and accountability from their governments. Grassroots organizations work tirelessly to build resilience in communities, advocate for policy changes, and spread awareness.

Countries are beginning to listen. Renewable energy adoption is increasing, electric vehicles are becoming more common, and cities are rethinking urban planning to reduce emissions and adapt to rising temperatures. Some of the world's largest corporations are pledging to go carbon-neutral, and governments are finally putting policies in place to phase out coal plants and support green energy infrastructure.

David Sinclair's **Lifespan** reminds us that humanity is capable of remarkable feats when it sets its mind to a task. The same ingenuity that led to the development of vaccines, extended human life expectancy, and sent us to the moon can be harnessed to combat the climate crisis. The story of global warming is not yet finished, and we have the tools to write a hopeful ending.

A Way Forward

The path forward requires unprecedented levels of cooperation, innovation, and determination. Countries like Guyana must be supported as they navigate the delicate balance between economic development and environmental responsibility. Wealthy nations must uphold their promises to fund green transitions, ensuring that the resources are available for those who need them most.

Developing nations can be the pioneers of a new kind of growth—one that learns from the mistakes of the past and seeks to build a future where prosperity and sustainability coexist. The Thwaites Glacier may be the wake-up call the world needs—a reminder that the stakes are higher than ever and that the time for action is now.

It's not too late to choose wisely. The Earth is calling for change. The question is: will we answer the call?

Planet in Our Hands:

Chapter 6:

The Technology of Today:
A Turning Point for Humanity

In our world today, technology is evolving at an unprecedented pace, bringing both incredible opportunities and serious risks. This chapter explores how advancements in artificial intelligence (AI), healthcare, food security, and biotechnology are shaping the future of humanity. Alongside these benefits, we must remain vigilant, for AI is not just a tool—it is the sixth horseman, following in the wake of climate change, the fifth horseman. These new forces challenge humanity's ability to control its own destiny, pushing us to harness technology responsibly.

The Dawn of a New Era

Picture a day where technology anticipates your every need. Your smart home adjusts the environment, makes your coffee, and manages your daily schedule seamlessly. For many people living in the developed world that day is here. AI-powered devices, which were once science fiction, now act as personal assistants, learning from your habits and ensuring comfort. Today, AI has become a silent force in everyday life, integrating into the fabric of society.

Planet in Our Hands:

The rise of AI is a fascinating journey. From its early days as a theoretical concept in computer science, it has evolved into a critical component across industries like healthcare, agriculture, and urban development. AI's ability to process massive datasets has made it invaluable in forecasting weather patterns, optimizing agricultural yields, and offering personalized medical treatments. It's as if humanity has unleashed a genie—one with the power to solve some of our most pressing global challenges.

A Revolution in Healthcare

One of the most transformative aspects of this technological era is its impact on healthcare. Diagnosing illnesses was once a time-consuming process involving numerous tests. Today, AI algorithms can analyze medical scans, genetic information, and patient records within seconds, offering accurate diagnoses faster than human doctors. Diseases like cancer, which used to require invasive procedures for early detection, are now identified in their earliest stages, vastly improving treatment outcomes.

CRISPR, a groundbreaking gene-editing technology, plays a crucial role here. Combined with AI, CRISPR isn't just about treating diseases; it's about curing them. By editing genetic material, scientists can correct the mutations responsible for conditions like cystic fibrosis and sickle cell anemia. This technology paves the way for a future where genetic diseases could be eradicated entirely. But there's more—CRISPR also brings the possibility of creating "superhumans." Imagine altering genes to extend life, enhance intelligence, or boost physical strength. This potential, while exciting, is ethically complex.

Without careful regulation, these powerful tools could be misused. The line between treating diseases and enhancing human capabilities is thin, and without ethical guidelines, we risk a future where genetic enhancement becomes accessible only to the wealthy, further widening societal inequalities.

Food Security in a Changing World

AI's influence extends beyond healthcare, transforming agriculture and food production. Traditionally, agriculture has depended on weather patterns and soil conditions. AI, however, paired with advanced sensors and satellite technology, is revolutionizing farming practices. AI systems analyze soil data, weather forecasts, and satellite images to deliver precise planting and harvesting schedules. This approach maximizes crop yields while minimizing water and fertilizer use, making agriculture more sustainable.

AI has also been instrumental in developing lab-grown meats and alternative protein sources. These innovations address the environmental impact of livestock farming, which contributes significantly to greenhouse gas emissions. With AI's guidance, scientists are engineering foods that are not only nutritious but also sustainable, reducing hunger in regions prone to food insecurity. This progress shows how technology can support humanity in feeding a growing population sustainably, offering hope against the backdrop of climate change—the 5th horseman.

The AI Horseman: Friend or Foe?

As AI grows in power, it becomes the sixth horseman—a force that, if not controlled, could shape the future unpredictably. AI is no longer limited to simple computations; it is capable of exploring pathways previously unimaginable to humans. One compelling example of AI's capabilities is its mastery of the ancient game **Go**. Unlike chess, **Go** is characterized by an almost infinite number of possible moves, making it a long-standing challenge for programmers. Traditional methods struggled to match the intuitive strategies employed by human players.

That changed with the introduction of AlphaGo, an AI developed by DeepMind. AlphaGo did not merely learn from past games; it taught itself through deep learning and self-play. During its match against the world champion, Lee Sedol, AlphaGo made moves that confounded experts—moves so unconventional they initially appeared nonsensical. Yet, these moves revealed a depth of understanding beyond human expertise, demonstrating that AI could explore avenues of 'thought' never before conceived.

This breakthrough highlighted AI's potential to innovate and solve problems in ways humans may not have imagined. It is now applied in

Planet in Our Hands:

fields like medicine, engineering, and climate modeling, where uncovering new strategies could lead to groundbreaking solutions. However, AlphaGo's success also serves as a cautionary tale: AI, with its capacity to think independently, can create unpredictable outcomes that, without proper regulation, could be dangerous.

Events in Myanmar illustrate the darker side of AI's power. Algorithms designed to maximize engagement on social media were manipulated to spread hate speech and incite violence against the Rohingya population. The genocide that followed was a stark reminder of how algorithms, when left unchecked, can be weaponized to divide and harm. This isn't an isolated incident—AI has also been deployed to manipulate elections, influence voter behavior, and disseminate misinformation. These algorithms lack a moral compass; they function based on the data and instructions they are given, prioritizing engagement even if it comes at the cost of truth and unity. This makes developing ethical standards and regulations essential to prevent AI from becoming a weapon of manipulation.

AI on the Battlefield

The application of AI in warfare shows another area where technology's potential for harm is evident. Autonomous drones and robotic soldiers are no longer confined to the realm of science fiction—they are being developed and tested globally. These machines can process battlefield data in real time, making decisions faster than any human. While this might reduce human casualties among soldiers, it also raises serious ethical and moral questions.

What happens when an AI, programmed to kill, makes a mistake? How do we ensure that autonomous weapons operate within the rules of war? Governments and international bodies are working to establish guidelines and agreements to control AI in warfare, but the rapid pace of technological development complicates these efforts. At the time of this writing AI is being tested in real time in real wars without agreed guidelines. There is always the risk that rogue states or organizations could deploy these technologies without any regard for human life or international law.

The Future of Work and Society

AI isn't just reshaping warfare and healthcare; it is also transforming economies and societies. Automated systems are increasingly taking over tasks that once required human labor, particularly in manufacturing and transportation. While this may lead to job displacement, it also creates new opportunities. As repetitive tasks become automated, people are freed to engage in more creative, intellectually stimulating work. As noted in the introduction, I've used AI to help me to write this book.

In education, AI personalizes learning experiences, tailoring lessons to each student's strengths and weaknesses. This revolutionizes education, making it more accessible and effective for people regardless of their background. With these advancements, AI has the potential to close educational gaps and provide the skills needed to thrive in an automated world.

Balancing Progress and Precaution

As humanity progresses, we must balance the excitement of technological breakthroughs with caution. AI holds the key to solving some of the world's most pressing challenges, from healthcare to climate change. It can optimize energy usage, reduce waste, and create sustainable solutions for a growing population. But as we've seen, unchecked AI has the potential to cause great harm.

Yuval Harari, in his book **Nexus**, calls AI "Alien Intelligence". This is appropriate because AI can 'think' in ways that humans never have before. Unchecked, it truly does have the ability to destroy us.

We stand at a crossroads. Will we use AI to build a more equitable, sustainable, and prosperous future for all, or will we allow it to deepen the divide between rich and poor, or worse, weaponize it against one another? The choices we make today will shape the future of our species.

Conclusion

The technological landscape of our era is one of promise and peril. AI, biotechnology, and other innovations have the power to uplift

Planet in Our Hands:

humanity, addressing some of our most significant challenges. But caution is advised. To secure a future where technology benefits all, we must establish ethical guidelines and regulations. Working together as a global community, we still can harness these tools wisely.

The genie is out of the bottle, and now it's up to humanity to guide it. The question remains: will we create a brighter, more equitable future, or will we let technology lead us down a darker path?

Chapter 7:

Misinformation and the Battle for Truth

In the information age, the promise of accessible knowledge has been overshadowed by the rise of misinformation—a force capable of reshaping perceptions, distorting facts, and influencing political and social outcomes. This chapter explores how misinformation has evolved, its impact on public health, political movements, and social stability, and how political leaders have harnessed it as a tool for power. It provides specific examples from industries, political campaigns, and the digital landscape to illustrate the pervasive and often destructive power of misinformation.

Tobacco - Manipulating Public Perception for Profit

The tobacco industry offers one of the earliest large-scale examples of misinformation designed to protect corporate profits. By the mid-20th century, scientific research had established a clear link between smoking and serious health risks, including lung cancer and heart disease. In response, the tobacco industry launched aggressive campaigns to obscure these findings.

Their strategy involved funding studies that questioned the health risks associated with smoking, publishing advertisements that portrayed smoking as glamorous or sophisticated, and framing the choice to smoke as a matter of personal freedom. Public relations firms played a key role in shaping the narrative, suggesting that the science was

Planet in Our Hands:

"uncertain" and that more research was needed. This approach was designed to create doubt among the public and delay regulatory measures that could hurt sales.

For decades, this misinformation campaign succeeded, allowing millions to continue smoking despite mounting evidence of its dangers. The industry's efforts prolonged its profitability at the cost of public health, resulting in millions of smoking-related deaths. The tobacco case demonstrates how misinformation, when backed by powerful interests, can significantly delay regulation and influence public behavior, even in the face of scientific consensus.

Sugar - Shaping Dietary Narratives

The sugar industry used similar tactics to influence public perception and shape dietary guidelines. As early as the 1960s, studies began linking sugar consumption to obesity, heart disease, and other health problems. In response, the sugar industry funded research that shifted the blame from sugar to dietary fats, influencing public health policies for decades.

This campaign was effective in reshaping dietary guidelines that encouraged low-fat, high-sugar diets. Products marketed as "low-fat" became associated with health, even though they often contained high levels of sugar. This misinformation had long-term effects, contributing to rising rates of obesity, diabetes, and other chronic health conditions worldwide. Today, the legacy of this campaign continues to affect public health and nutrition, illustrating the power of strategically manipulated information.

Climate Change – Denying a Great Threat

The denial of climate change and global warming offers another stark example of how information can be manipulated to serve powerful interests. Despite overwhelming scientific consensus on the causes and dangers of climate change, industries tied to fossil fuels have mounted extensive misinformation campaigns to create doubt and delay regulatory action. Borrowing tactics from the tobacco and sugar industries, these campaigns fund studies that downplay the effects of greenhouse gases and promote the idea that climate change is part of a natural cycle rather than a result of human activity. Through targeted messaging, these organizations frame environmental regulations as threats to economic stability, rallying support by appealing to fears about job loss and economic decline.

Social media amplifies these narratives, creating echo chambers where misinformation about climate science thrives, often portraying climate change as a "debate" rather than a well-established reality. By distorting the facts, these campaigns have succeeded in delaying meaningful action, exacerbating the climate crisis and compromising public trust in science. Unfortunately, too many people now believe that climate change is a "hoax", instead of the threat to our continued existence that it truly is.

Brexit - Misinformation in Political Campaigns

Misinformation is not limited to public health or global warming; it plays a powerful role in political campaigns. The 2016 Brexit referendum in the United Kingdom serves as a modern example of how falsehoods can influence national decisions. Brexit proponents framed the European Union as a bureaucratic force draining British resources and compromising national sovereignty.

One of the most notorious falsehoods during the campaign was the claim that leaving the EU would free up significant funds for the National Health Service (NHS). This, despite being widely debunked by experts, was featured prominently on campaign materials, appealing to voters who felt shortchanged by the EU. The campaign also used fearmongering around immigration, suggesting that remaining in the EU would lead to uncontrolled immigration levels and economic strains.

Social media played a crucial role in spreading these narratives. Targeted ads, often containing misleading or false information, reached millions of undecided voters. Algorithms designed to maximize engagement amplified emotionally charged content, further polarizing the public. In the end, misinformation played a significant role in the narrow decision to leave the EU, a choice that continues to have profound economic and political consequences.

Planet in Our Hands:

Political Leaders - The Use of AI and Misinformation in Elections

Around the world, political leaders have increasingly used misinformation and AI technology as tools to gain and consolidate power. These tactics often involve bypassing traditional media channels in favor of direct communication through social media, which allows for unfiltered messaging that can reach and influence millions of voters.

In several countries, political leaders have exploited AI-driven technology to target specific demographics with tailored messages designed to manipulate opinions and deceive voters. For example, campaigns have used AI algorithms to analyze social media behavior and identify issues that resonate emotionally with voters. They then create targeted ads that amplify these sentiments, often through misleading or false narratives. This method has been effective in mobilizing support by appealing directly to voters' emotions and fears.

In some regions, AI-powered bots flood social media with messages designed to create the illusion of widespread support or opposition. These automated accounts amplify misinformation, making it appear as though certain narratives or viewpoints have significant public backing, even when they do not. This tactic was used in various elections to create echo chambers where misinformation thrives, reinforcing the beliefs of target audiences and drowning out contradictory or fact-based information.

Another approach involves spreading falsehoods about political opponents. In recent elections, misinformation campaigns have used AI-generated deepfakes—realistic but entirely fabricated videos—to discredit rivals. These videos are crafted to appear authentic, showing opponents making controversial statements or engaging in activities that never occurred. The goal is to sway undecided voters by presenting manipulated content as evidence of an opponent's untrustworthiness or incompetence.

In other instances, political leaders have spread misinformation about public institutions, such as the electoral process or the judiciary, to undermine trust and position themselves as the only credible source of truth. By casting doubt on the reliability of democratic systems, they rally supporters around the idea that the system is rigged against them. This tactic not only mobilizes support but also delegitimizes institutions designed to provide checks and balances, consolidating power in the hands of these leaders. Consider a recent election in your own country. Ask yourself if your vote might have been influenced by unsubstantiated information that you'd gleaned from politically influenced social media.

The Role of Social Media - Amplifying Misinformation

Social media platforms have become powerful tools for spreading misinformation, particularly in political contexts. Platforms like Facebook, X (formally known as Twitter), and YouTube prioritize content that maximizes user engagement, often promoting sensational or emotionally charged posts. This makes social media an ideal platform for misinformation, as false or divisive stories generate high levels of interaction.

The algorithms used by these platforms amplify content that provokes strong reactions, creating echo chambers where users are predominantly exposed to information that reinforces their pre-existing beliefs. Political leaders and their allies exploit these echo chambers to disseminate targeted messages, mobilize supporters, and undermine their opponents.

In some regions, the consequences of social media misinformation have been severe. Misinformation campaigns have been used to incite violence or create fear among the public. For instance, groups have used social media to spread falsehoods about minority communities, portraying them as threats to national security. Algorithms that prioritize engaging content amplify these messages, leading to real-world violence and social unrest. This illustrates how social media, when unregulated, can become a powerful tool for spreading harmful misinformation.

Artificial Intelligence and Misinformation - The New Frontier

Artificial intelligence (AI) introduces new complexities in the battle against misinformation. Technologies like deepfakes, which use AI to create realistic but fabricated videos and audio recordings, make it increasingly difficult to distinguish between real and manipulated

Planet in Our Hands:

content. This technology can be used to fabricate evidence, discredit political opponents, and manipulate public perception at an unprecedented scale.

AI algorithms already influence much of the content seen on social media. Automated bots, powered by AI, generate and amplify posts designed to spread false information and promote divisive narratives. During elections in various countries, AI-powered bots have been deployed to manipulate public opinion, creating the illusion of widespread support for certain political causes and discrediting opponents.

The potential for AI-driven misinformation to destabilize societies is significant. As AI continues to develop, the ability to blur the line between truth and fiction becomes easier, posing significant challenges to maintaining an informed public.

Strategies for Navigating the Misinformation Landscape

To navigate an environment saturated with misinformation, individuals need to develop critical thinking skills and use a range of strategies to assess the credibility of information. These strategies include:

1. **Diversify News Sources**: Consuming information from various reputable sources, such as BBC, Al Jazeera, Reuters, The Guardian and The Associated Press, provides a broader and more balanced view of events. This approach helps avoid the pitfalls of echo chambers that reinforce pre-existing biases.

2. **Verify Information**: Fact-checking organizations like Snopes, PolitiFact, and FactCheck.org are valuable resources for verifying claims made in the media or on social platforms.

3. **Understand Media Bias**: Recognizing that all media outlets have some level of bias helps individuals critically evaluate the information they receive. Knowing the political leanings of different outlets can contextualize coverage and highlight areas where further scrutiny may be needed.

4. **Promote Media Literacy**: Education in media literacy is essential for identifying misinformation. Schools and community programs should

teach people how to evaluate the credibility of sources, recognize bias, and understand the role of algorithms in shaping information.

5. **Be Skeptical of Viral Content**: Stories that provoke strong emotional reactions are often designed to manipulate or mislead. Verifying the authenticity of such content through multiple reliable sources is crucial before accepting or sharing it.

6. **Support Independent Journalism**: Independent and nonprofit journalism organizations provide investigative reporting without the influence of corporate or political interests. Supporting these outlets helps sustain credible and in-depth journalism.

Conclusion: Confronting Misinformation

Misinformation poses one of the most significant threats to modern society, shaping political outcomes, influencing public health decisions, and undermining social stability. Political leaders and powerful interest groups use it as a tool to achieve their goals, often at the expense of public trust and societal cohesion. The rise of advanced technologies like AI only amplifies this challenge, making it more difficult to discern truth from fiction. In this complex landscape, critical thinking, diverse information sources, and support for independent journalism remain essential for preserving the integrity of information and democracy.

Planet in Our Hands:

52

Chapter 8:

What We Have Learned and The Moment of Reckoning

In the grand expanse of the universe, Earth is but a speck—a fragile, blue marble suspended in a vast cosmic sea. For billions of years, this planet has spun gracefully around its star, enduring asteroids, ice ages, volcanic eruptions, and the shifting of continents. Life has adapted, evolved, and transformed through every twist and turn, from simple bacteria to towering dinosaurs, and eventually, to us—humans. In the span of Earth's 4.5 billion years, humans have occupied a mere fraction of time, but in that short period, we've become the most influential species ever to walk its surface.

Our story began like a spark, millions of years ago, when our ancestors first stood upright and gazed upon the horizon. In **Sapiens**, Yuval Harari explains how that simple shift in posture set us on a path unlike any other. We learned to harness fire, create tools, and communicate through complex language. We crossed continents, sailed oceans, and built civilizations. We discovered the power of science, unlocked the secrets of our genes with technologies like CRISPR, and reached beyond our planet, exploring the cosmos with telescopes and probes.

Yet, as remarkable as our achievements are, they come with a price. We are now at a pivotal moment—a tipping point where our mastery over nature has begun to threaten our own existence.

Planet in Our Hands:

The Good We've Done

Before we look at the shadows of our journey, it's important to acknowledge the light we've brought into the world. The story of human progress is a tale of remarkable accomplishments. Advances in medicine, such as vaccines and antibiotics, have saved countless lives. According to David Sinclair's **Lifespan**, we've extended human life expectancy by decades, pushing back the boundaries of aging itself. Biotechnology, like the CRISPR gene-editing technology discussed by Jennifer Doudna in **A Crack in Creation**, offers the promise of eliminating genetic diseases and enhancing human capabilities.

In the 20th and 21st centuries, we saw revolutions in technology that connected the globe, allowing us to share knowledge at the speed of light. AI and machine learning have transformed industries, improved diagnostics in healthcare, and even aided in disaster response. Bill Gates, in **How to Avoid a Climate Disaster**, emphasizes the progress we've made in clean energy technology, creating solutions that could potentially steer us away from the brink of environmental catastrophe.

We have fought against the ancient horsemen of war, famine, disease, and death. In many ways, we've won. Global conflicts have diminished in scale, with international institutions promoting peace. Agriculture has advanced, producing food on a scale our ancestors could never have imagined. Diseases that once ravaged entire populations have been eradicated or controlled, thanks to scientific breakthroughs and vaccines. It is clear that we, as a species, have the capability to create a world where suffering is minimized, and prosperity is within reach for all.

The Shadow We've Cast

But with every light, there is a shadow. While we've created marvels, we've also laid the foundations for potential collapse. The Earth, resilient as it is, has limits. And we are testing those limits at an unprecedented pace.

In **The Sixth Extinction**, Elizabeth Kolbert paints a sobering picture of the Anthropocene—the age of humans. We have altered the planet's ecosystems so drastically that species are vanishing at a rate not seen since the last mass extinction, 65 million years ago. Our thirst for resources, driven by expanding industries and populations, has led to deforestation, pollution, and the acidification of oceans.

Climate change, the fifth horseman, is the most pressing and visible consequence of our influence. As Jeff Goodell describes in **The Heat Will Kill You First**, our planet is warming faster than at any point in human history. Heatwaves, hurricanes, and rising sea levels are already impacting millions, with developing countries bearing the brunt. The evidence is undeniable: the ice caps are melting, the oceans are rising, and weather patterns are becoming more extreme. Katharine Hayhoe, in **Saving Us**, highlights how these changes are not just environmental issues; they are social issues, affecting vulnerable communities first and hardest.

And while Earth itself has survived past climate shifts the difference now is that the very systems we rely on—agriculture, cities, economies—are not designed to withstand such rapid changes. The irony is that while we have the tools to combat climate change, our political and economic systems often delay action, prioritizing short-term gains over long-term survival.

Misinformation and the New Horsemen

Misinformation has become a powerful tool, spreading at an alarming rate through social media and other digital platforms. False narratives about climate change, public health, and politics have sown division and confusion. Except that this time the power of AI is also in play. In **Nexus**, Harari warns about the dangers of this fragmented reality where facts become subjective, manipulated by those seeking power or profit.

The similarities between the communication strategies of modern populist leaders and those used by past dictators are striking. Hitler's ability to manipulate emotions and rally people to his cause is not far removed from the tactics employed by contemporary figures. The constant barrage of misinformation has led many to deny scientific realities—such as the harmful effects of cigarettes, the dangers of excessive sugar intake, and the urgency of the climate crisis. As **The Coming Wave** by Mustafa Suleyman suggests, the unchecked power of

Planet in Our Hands:

artificial intelligence, the sixth horseman, could worsen this, creating deepfake technology and sophisticated algorithms that blur the line between truth and falsehood.

The AI Dilemma

Artificial intelligence, which I've called the sixth horseman, holds both promise and peril. It has the potential to solve problems in ways that were once unimaginable—to aid in medical research, optimize energy usage, and even develop new ways to combat climate change. Yet, its rapid development raises critical questions about control, ethics, and the potential for abuse. If left unchecked, artificial intelligence could widen the gap between truth and fiction, manipulate public opinion, or even wage wars more efficiently.

Artificial intelligence's power to influence and predict behavior also brings up concerns about privacy and autonomy. As Suleyman outlines, we are at risk of losing control over these technologies, which, if misused, could serve as tools for authoritarian regimes or corporations prioritizing profit over human welfare. It's a paradox: the very technology that could help us overcome the challenges we face might also amplify them, pushing us further into conflict and chaos.

The Planet Will Survive—But Will We?

Throughout Earth's long history, it has faced cataclysmic events, from asteroid impacts to massive volcanic eruptions. Each time, life has rebounded, evolving and adapting in new and unexpected ways. But it is not Earth's survival that is at stake—it is ours.

Our actions have made our own environment hostile. Rising temperatures threaten to make parts of the planet uninhabitable. The choices of countries like Guyana, where newfound oil reserves offer both economic opportunity and environmental risk, highlight the challenges of balancing development and sustainability. If humanity continues on its current path, ignoring the signs and refusing to take collective action, we risk becoming the architects of our own extinction.

The Earth will endure, reshaping itself in the face of whatever we do. It has done so before, and it will do so again. But the question remains: will we be a part of that future?

The Choice Before Us

The good news is that it's not too late. Our story is still being written, and we have the power to change its course. By embracing the tools we've created—technology, science, and cooperation—we can choose a path that sustains life and ensures a future for generations to come. We must be vigilant, however, of the new horseman we have unleashed, ensuring that artificial intelligence serves humanity rather than undermining it, and that misinformation is replaced by a shared commitment to truth.

We are on the edge of a new chapter. The next pages could tell the story of a civilization that rises to meet its greatest challenges, or they could recount a cautionary tale of what happens when a species wields more power than it understands.

The choice is ours.

Planet in Our Hands:

Chapter 9:

The Power of One in a Changing World

In a world where the stakes have never been higher, it's easy to feel overwhelmed by the enormity of the challenges we face. Climate change, misinformation, and the rapid evolution of artificial intelligence seem like distant forces beyond our control. So, what are we going to do with all that we have learned so far. These last three chapters are action chapters. In this chapter we will look at things that individuals can do, on their own, to turn things around. In chapter 10 we will look at some ideas for groups or for individuals who would like to be more active, either in communities, their countries or even on a global scale. Chapter 11 ends this book with a dream of what we can achieve over the next 100 years, if we start to work now. 100 years may seem like a long time, but it sets a realistic goal given all the challenges that we face today. Truthfully, it is not even that far away. Given the promises of our medical health sciences, it is very possible that many persons who are 25 years old or younger in 2024 will live to see the year 2125.

We, as individuals, have the power to make real, tangible changes. The decisions we make in our everyday lives, from the food we eat to how we move around, can help shift the balance. It all starts with the choices we make, every single day.

Planet in Our Hands:

The Choices We Make: Eating Less Meat

The first step an individual can take is to reduce their meat consumption. This is not just about adopting a new diet trend; it's a powerful way to reduce greenhouse gas emissions. The livestock industry is one of the biggest contributors to methane and carbon dioxide emissions. Methane, which is released during the digestive process of ruminant animals like cows, is significantly more potent than carbon dioxide in trapping heat in the atmosphere.

To put it in perspective, the meat industry is responsible for nearly 15% of global greenhouse gas emissions—more than the entire transportation sector combined. Livestock farming also requires vast amounts of land and water, driving deforestation and the destruction of habitats. By eating less meat, particularly beef, individuals can dramatically reduce their carbon footprint. Remember that beef shows up in many forms – including hamburgers, maybe hot dogs and often in sandwiches.

Shifting to plant-based meals a few times a week can make a difference. It's about progress, not perfection. Imagine millions of people choosing to eat just one less meat-based meal per week. The cumulative impact would be monumental, creating a ripple effect that could lead to less land being cleared, fewer animals being raised under industrial conditions, and less methane being released into the atmosphere.

Driving Change: Electric, Hybrid, and Hydrogen Cars

The vehicles we drive are another major contributor to greenhouse gas emissions. Passenger vehicles alone are responsible for nearly a fifth of global CO2 emissions. Switching to electric, hybrid, or hydrogen-powered cars can make a significant difference. Advances in technology mean that used electric cars are now much closer in price to their gasoline counterparts, making them a more accessible option for many people.

Electric vehicles (EVs) produce zero emissions at the point of use, but it's important to consider the full lifecycle. When charged with electricity from clean, renewable sources like wind, solar, nuclear or hydro, EVs dramatically reduce carbon emissions compared to traditional gas-powered cars. Even when charged using a grid that includes fossil fuels, they still emit fewer greenhouse gases over their lifespan.

In addition to their environmental benefits, EVs are also much cheaper to maintain. With fewer moving parts than traditional vehicles, there are no engine components that require regular maintenance, like oil changes, spark plugs, or exhaust systems. This results in lower repair costs and fewer visits to the mechanic, making them not only a greener but also a more economical choice over the long term.

For those who aren't in the market for a new car, there are other ways to minimize emissions: carpooling, using public transportation, and bicycling are all effective ways to cut down on carbon emissions. By reducing the number of single-occupancy vehicles on the road, we also reduce traffic congestion, which further decreases emissions and improves air quality in our cities.

A Note on Carbon Credits

Another option for individuals looking to minimize their carbon footprint is purchasing carbon credits. Carbon credits are permits that allow a person or organization to emit a certain amount of carbon dioxide. Each credit typically represents one ton of CO_2. By purchasing these credits, individuals and companies can offset their emissions, as the money from these purchases is used to fund projects that reduce or capture emissions, such as reforestation initiatives, renewable energy development, and methane capture from landfills.

While buying carbon credits doesn't eliminate emissions directly, it helps fund the transition to a cleaner, greener future. It's a way of taking responsibility for unavoidable emissions, especially for those who may not yet have access to clean technology options. However, it's important to purchase from verified programs that ensure the credits genuinely contribute to sustainable projects. Consider buying carbon credits when you travel by airplane or take a cruise vacation.

Rethinking Waste: Recycling, Reusing, and Reducing

Recycling has been promoted as a way to reduce waste, but it's not a silver bullet. Many places lack the infrastructure for comprehensive recycling, and much of what we believe is recycled often ends up in landfills. That's why reusing, repurposing, and reducing consumption—especially of single-use plastics—are more effective strategies.

Single-use plastics, from water bottles to packaging, create enormous waste problems. They don't biodegrade; instead, they break down into microplastics that pollute oceans and harm wildlife. By choosing reusable products—water bottles, shopping bags, containers—we minimize waste and decrease the demand for single-use items.

Consider donating used household items and clothes to agencies or businesses that will resell or repurpose them. Consider buying your own clothes from vintage stores.

Composting organic waste is another step that can be taken at home. Food scraps and yard waste that end up in landfills release methane as they decompose. By composting, we can turn organic waste into nutrient-rich soil for gardens and plants, reducing methane emissions and enriching our environment.

Energy Use: Lighting, Heating, and Beyond

Reducing fossil fuel consumption in our homes is another critical step. Switching to energy-efficient LED lighting and using clean electric energy for cooking and heating, if available, are practical ways to minimize our impact. LEDs use up to 80% less energy than traditional incandescent bulbs and last significantly longer, reducing both energy consumption and waste.

In regions where clean electricity is available, transitioning from fossil fuel-based heating to electric options—like heat pumps—further reduces carbon emissions. For those in areas where access to clean energy is limited, investing in solar panels or other renewable energy sources can help make a household more sustainable.

Conserving Water: A Precious Resource

Water conservation is not just about saving money on utility bills; it's about protecting a precious and limited resource. Freshwater availability is decreasing in many parts of the world due to climate change, overuse, and pollution. Simple steps like fixing leaks, taking shorter showers, turning off the tap while brushing your teeth and using water-efficient appliances can reduce household water use.

Growing drought-resistant plants and landscaping with native species can also save water. In agriculture, using techniques like drip irrigation conserves water and maximizes efficiency. Whether it's on a large scale or at home, every drop counts.

Eating and Shopping Locally: Supporting Sustainable Choices

Buying locally produced goods and growing your own food are powerful ways to reduce your carbon footprint. Food transportation accounts for a significant portion of global emissions; eating food grown close to home reduces the energy required for shipping and distribution.

Local farmers' markets often provide fresh, organic produce that supports small-scale, sustainable agriculture. For those with the space, starting a small garden can supplement food needs and reduce the environmental impact of large-scale farming. Growing herbs, vegetables, or fruit trees can also be a rewarding and sustainable way to connect with the earth.

Choosing Leaders Who Prioritize Climate Action

One of the most impactful decisions we make as individuals is choosing leaders who are genuinely committed to solving the climate crisis. The policies and decisions made by elected officials have far-reaching effects on environmental regulations, clean energy initiatives, and sustainability measures. When voting, look for candidates with clear, actionable plans to address climate change, invest in renewable energy, and protect our natural resources. Leaders who prioritize environmental policies help set the stage for lasting change on a scale that individual efforts alone cannot achieve.

Planet in Our Hands:

Checklist: Steps You Can Take Today

- **Eat Less Meat**: Incorporate more plant-based meals into your diet. Consider making every Monday a 'Meatless Monday'.
- **Drive Cleaner**: If possible, choose electric, hybrid, or hydrogen cars, or ride a bicycle, use public transport or share transport, for example carpooling or ride sharing.
- **Purchase Carbon Credits**: Offset unavoidable emissions by investing in verified carbon credit programs.
- **Recycle, Reuse, Reduce**: Cut down on single-use plastics, and focus on buying items that can be repurposed after you've used them and if, safe and practical, on buying items that have been used by others. Consider habits that will keep products out of landfills.
- **Use Energy Wisely**: Switch to LED lighting and, where possible, clean electric heating and cooking.
- **Conserve Water**: Shorten showers, fix leaks, and use water-efficient appliances.
- **Buy Local**: Support local markets or grow your own food to minimize transportation emissions.
- **Vote for Climate**: Support leaders with clear, actionable climate policies. Avoid leaders who deny climate change.

By taking these steps, we create a ripple effect, not just reducing our own carbon footprint but inspiring those around us to do the same. The power of one multiplies when shared across communities, nations, and eventually, the world.

So, reader, before you go on to the next chapter, pause for a minute and answer this question. What are the one or two things that you are going to do, right now, towards solving the problems that we face? We have just shared eight ideas for you to consider. Before doing anything else, get a pen and paper and write down one new thing that you are going to do starting right now. Even the smallest action will make a difference. If every reader of this book took just one small step, then humanity will take a leap forward. Think about it. What is your first step going to be?

Now, go and do it.

Chapter 10:

Collective Action: Uniting for Change

While individual actions are critical, the challenges humanity faces require collective efforts that extend beyond personal choices. History has shown us that, despite our differences, people can come together in remarkable ways to address shared threats. By pooling resources, knowledge, and energy, we can create powerful movements capable of bringing about substantial change. The battle against climate change, the fight to combat misinformation, and the campaign to harness technology for the collective good—these all demand a unified, strategic approach. It is in this spirit of unity that we can build a more sustainable and just future for everyone. Here's how communities and societies can come together to turn the tide.

Avoiding Wars and Conflicts: A Call for Unity

Wars are often driven by greed, the pursuit of power, and the desire for control over resources. Throughout history, despots, tyrants, and populist politicians have used these motives to incite conflict, leaving behind a legacy of destruction and loss. These leaders thrive by exploiting societal divisions, manipulating fears, and stoking nationalism to gain or maintain power. The past century alone is filled with examples where millions of lives were lost in wars that could have been averted through diplomatic negotiations, international cooperation, and a commitment to peace.

Planet in Our Hands:

To prevent future conflicts, we must reject the divisive rhetoric that feeds on fear and separates people into "us" versus "them." This requires us, as citizens, to remain vigilant and critical of those who seek to manipulate emotions for political gain. Voting for leaders who prioritize peace, diplomacy, and the well-being of their people over aggression and division is one of the most powerful ways we can contribute to a peaceful world.

If you would like to play a more activist role, or even become politically involved in fighting for change then here are some specific points for you to lobby governments or for political aspirants to campaign on.

- **Peacebuilding and Diplomatic Programs**: Advocate for greater investment in international diplomacy and peacebuilding missions. Citizens should push their governments to actively participate in multilateral peace initiatives, such as those organized by the United Nations or other regional alliances, to prevent conflicts before they escalate. Political aspirants can build their platforms around enhancing diplomatic ties, supporting negotiations over military intervention, and fostering international dialogue to solve disputes.

 - **Restricting and Regulating Arms Sales**: Lobby for stricter laws regulating the sale and distribution of weapons, particularly to regimes with poor human rights records or to countries on the brink of conflict. Political candidates should prioritize creating transparent and responsible arms trade policies, ensuring that weapons do not fall into the hands of those likely to use them for harm. A campaign focused on disarmament and responsible defense strategies can appeal to voters who desire long-term global stability.

- **International Cooperation Agreements**: Support politicians who emphasize the importance of international cooperation, especially in fields like climate change, humanitarian aid, and conflict resolution. Campaigns can highlight the benefits of strengthening international bodies that promote peace, such as the International Criminal Court, and advocate for policies that hold leaders accountable for war crimes. By promoting a vision of a united international community working for peace, political leaders can build trust and engagement with voters.

It is only by collectively rejecting warmongering and embracing peace-building initiatives that we can create a world where cooperation and understanding become the norm, replacing the cycle of violence and destruction. Educating ourselves about the motivations behind conflicts, questioning leaders who use aggressive rhetoric, and organizing peaceful demonstrations are all ways we, as communities, can act in unity to avoid the horrors of war.

Combatting Misinformation: Finding the Truth in a Noisy World

In today's digital age, misinformation spreads faster and more effectively than ever before. Fueled by social media algorithms designed to maximize engagement rather than ensure accuracy, false information circulates widely, shaping opinions and influencing behaviors in dangerous ways. Misinformation has the power to sway elections, undermine trust in science, and distort public understanding of critical issues like health and climate change. This is not a minor problem; it is a global crisis that threatens our ability to make informed, rational decisions.

Populist politicians and manipulative figures capitalize on this chaos, using misinformation to stoke fear and confusion among the public. The rise of these leaders is often tied to their ability to create narratives that play on people's anxieties, casting doubt on established facts and institutions. To counteract this trend, communities must become proactive in seeking the truth and supporting measures that curb the spread of false information.

Specific points to lobby governments or for political aspirants to campaign on include:

- **Strengthening Fact-Checking and Media Literacy Standards**: Advocate for government support of independent fact-checking organizations and the establishment of national standards for media literacy education. Communities should pressure their governments to implement media literacy curricula in schools, ensuring that future generations are equipped with the tools to critically evaluate information. Political aspirants can campaign on the importance of

Planet in Our Hands:

educating the public, not only to combat misinformation but also to foster an informed and engaged electorate.

- **Regulating Social Media Platforms**: Push for policies that hold social media companies accountable for the dissemination of false information. Governments should enforce regulations that require these platforms to be transparent about how their algorithms work and to take active steps in limiting the reach of misinformation. Aspirants can focus their campaigns on developing policies that mandate these platforms to flag misleading content, provide reliable information, and limit the influence of fake news, especially during critical periods like elections.

- **Funding Public and Independent Media**: Lobby for increased funding for public broadcasting services and independent media outlets that prioritize unbiased, factual reporting. By ensuring these outlets have the resources to provide accurate and timely news, governments can offer citizens reliable alternatives to the often-biased and sensationalized content found on social media. Candidates can campaign on the need for accessible, high-quality journalism that reaches all communities, emphasizing the importance of protecting freedom of the press while maintaining credibility and accountability.

Educating communities about media literacy and organizing events that promote critical thinking are ways in which societies can fight back against the tide of misinformation. By demanding transparency, we not only protect our democratic institutions but also ensure that decision-making is rooted in truth and not in manipulation or deceit.

Climate Action: Collective Steps to Secure Our Future

Climate change represents one of the most complex and urgent threats facing humanity. The scale of its impact demands a coordinated global response. Governments, businesses, and citizens must work together to reduce emissions, protect ecosystems, and transition to renewable energy sources. This coordinated effort requires clear, actionable policies, massive investments in green technology, and a collective willingness to adapt to a new, sustainable way of life.

Specific points to lobby governments or for political aspirants to campaign on include:

- **Implementing Carbon Pricing Mechanisms**: Advocate for carbon pricing systems that tax emissions, providing an economic incentive for businesses to reduce their environmental impact. Carbon pricing not only holds polluters accountable but also generates revenue that can be reinvested in clean energy projects and support for communities transitioning away from fossil fuels. Political candidates can campaign on implementing fair carbon pricing policies that ensure accountability while funding sustainable development programs. Populist politicians find carbon pricing to be an easy target. No one likes to pay more taxes, even if they eventually get it back. Taxing carbon pollution is not the only answer, but it is a very credible one.

- **Incentivizing Renewable Energy**: Push for tax incentives and grants for businesses and homeowners to invest in solar panels, wind turbines, and other renewable energy sources. Communities should advocate for legislation that promotes building infrastructure for clean energy and the creation of jobs in the renewable sector. Candidates can present platforms that emphasize not only the environmental benefits but also the economic opportunities tied to the green energy transition, appealing to voters interested in sustainable growth and job creation.

- **Protecting and Restoring Natural Ecosystems**: Lobby for policies aimed at protecting vital ecosystems like forests, wetlands, and marine environments, which play critical roles in carbon absorption and biodiversity preservation. Aspirants can build campaigns that emphasize the importance of reforestation programs, conservation efforts, and the sustainable management of natural resources. By promoting these initiatives, political leaders can highlight the interconnectedness of environmental health and human well-being, appealing to citizens who value long-term ecological balance.

Building a Fair and Inclusive World: Acceptance and Climate Refugees

As the climate crisis intensifies, so does the displacement of people. Extreme weather, rising sea levels, and environmental degradation force

Planet in Our Hands:

millions to leave their homes. To ensure the survival and dignity of these individuals, we must adopt policies and attitudes that embrace inclusion, providing climate refugees with the support and opportunities they need to rebuild their lives. This is a humanitarian issue that demands immediate attention and collective effort.

Specific points to lobby governments or for political aspirants to campaign on include:

- **Establishing Pathways and Legal Protections for Climate Refugees**: Advocate for policies that create legal pathways for climate refugees to resettle safely. Governments must be pressured to develop frameworks that provide climate refugees with access to housing, healthcare, education, and employment. Many people feel that climate refugees will all be from 'other countries'. The example of Isle de Jean Charles in the USA is a clear example that this is not so. If the Thwaites Glacier melts, the United States will be faced with millions of local climate refugees. Political candidates can campaign on ensuring that countries not only accept refugees but also actively integrate them into their communities.

- **Supporting Vulnerable Nations through International Aid**: Encourage international cooperation to support nations that are particularly vulnerable to the effects of climate change. This includes funding sustainable development projects, infrastructure adaptation programs, and technological aid. Aspirants can promote global solidarity by highlighting international agreements that commit financial and technological support to countries at risk, positioning themselves as champions of both humanitarianism and global responsibility.

- **Promoting Inclusion and Equal Opportunities**: Push for domestic policies that guarantee fair treatment and opportunities for all, regardless of refugee status, background, or ethnicity. Candidates should emphasize platforms that build inclusive communities, advocating for equal access to education, healthcare, and employment for everyone. Promoting diversity not only strengthens communities but also enhances resilience in facing climate challenges.

Acceptance extends beyond refugees. It also means promoting equality for all, building societies that value diversity and provide opportunities for every individual. A fair and inclusive society is one that is better prepared to tackle the collective challenges we face.

Checklist: Collective Actions for a Sustainable Future

- **Promote Peace**: Advocate for policies that prioritize diplomacy over conflict and support international cooperation for conflict resolution.

- **Fight Misinformation**: Lobby for regulations that hold social media platforms accountable and promote national standards for media literacy.

- **Demand Climate Action**: Support carbon pricing, renewable energy incentives, and policies protecting ecosystems.

- **Welcome Climate Refugees**: Campaign for pathways that provide refugees with access to essential resources and advocate for global aid.

- **Foster Equality**: Promote fair treatment, diversity, and inclusion within communities to strengthen social resilience.

By uniting around these principles and advocating for policies that reflect these values, communities can lead the way in creating a sustainable, peaceful, and equitable world. When the collective will of the people is mobilized, it becomes an unstoppable force capable of reshaping the future.

Planet in Our Hands:

Chapter 11:

The 100 Year Mission: A Vision for a Reimagined Future

Imagine standing at the edge of a vast landscape, looking out at a world that could be—**a world transformed by the choices we make today**. It's hard to picture the distant future, a hundred years from now, when our bones have long since turned to dust and our voices are echoes in history. But some of us alive today may still be here to witness the end of this grand mission. Science and technology have stretched the boundaries of human longevity, and the possibility exists that the children born today, or even some adults among us, might live long enough to see this vision unfold. What if we could carve a path so clear, so resolute, that the people of tomorrow would look back and know we cared enough to change course? What if we could set in motion a mission that stretches across generations, building a bridge between the world we've inherited and the one we want to create?

This is **The 100 Year Mission**. It's not just a plan; it's a blueprint for humanity's survival and evolution. It's a promise that our collective actions will not only sustain us but also transform the way we live on this planet, bringing a sense of balance, justice, and sustainability to every corner of the Earth. It's our plan to overcome the great adversaries—the four horsemen that have plagued humanity since time immemorial: war, famine, disease, and death. And now, we must also face the two new threats of our own making: climate change and

Planet in Our Hands:

artificial intelligence. Let's walk through this vision, decade by decade, and see what each step means for us and those who will come after.

The First Decade: Laying the Foundations (Years 1-10)

The journey begins with a bold declaration: the time for hesitation is over. The next ten years are about laying the groundwork for the changes to come. We begin with education and awareness. The people must know the stakes—we're not just fighting to save the polar bears or the rainforests; we're fighting for our survival and the survival of every child born from this point onward. And more than that, we are beginning the long process of dismantling the horsemen who threaten our future.

Key Actions:

- **Universal Climate Education**: Implement global education programs that focus on climate change, sustainability, and the responsible use of technology. It starts in schools but extends into communities, workplaces, and governments. By ensuring that every child learns the truth about climate change and how to combat it, we set the stage for an informed, engaged populace capable of challenging the forces of misinformation.

- **Massive Renewable Energy Investments**: Governments, corporations, and individuals must invest heavily in renewable energy. Wind farms, solar panels, and hydroelectric plants need to become the norm, not the exception. Incentives should be in place for families and businesses to switch to green energy, making it accessible and affordable for everyone. In doing so, we take steps to combat the horseman of famine, reducing the environmental degradation that threatens food production worldwide.

- **Transition to Sustainable Agriculture**: Transform the way we produce food by moving away from industrial farming methods that deplete the soil and contribute to greenhouse gas emissions. Support regenerative agriculture practices that restore land, capture carbon, and sustain biodiversity. Farmers need access to new technologies and training to make this transition smoothly, ensuring food security and attacking the root causes of hunger.

By the end of the first decade, the seeds of transformation will be planted. The air will be cleaner, and renewable energy will light up more homes than ever before. There will be a growing sense that we are no longer simply reacting to crises but are actively shaping a new path.

The Second Decade: Transformation and Innovation (Years 11-20)

The foundations laid, the next decade is all about accelerating transformation and leveraging technology for good. It's here that we double down on our efforts, pushing beyond incremental progress to begin seeing tangible, widespread change. We also confront disease, that ancient scourge, by ensuring healthcare reaches every corner of the world.

Key Actions:

- **Electrification of Transport Systems**: We electrify our cities. Public transport systems convert to electric buses, trains, and trams, powered by clean energy grids. Incentives for electric vehicles expand, making them the dominant choice in urban and rural areas alike. Hydrogen fuel technology, too, finds its place, especially in long-haul transportation, creating a landscape where the hum of electric engines replaces the roar of fossil-fueled machines.

- **Reforestation and Restoration Programs**: On a global scale, forests lost to deforestation are replanted and protected. Wetlands are restored, and coastal areas once stripped of mangroves are revived. These areas not only absorb carbon but also provide habitats for wildlife, safeguarding biodiversity. Communities are encouraged and financially supported to participate in these efforts, understanding that they are the guardians of their environment.

- **Global Healthcare Access and Pandemic Preparedness**: We create a healthcare system that extends beyond national borders, preparing humanity to prevent and respond to pandemics with swift, coordinated action. Investment in medical research, vaccines, and healthcare infrastructure becomes a priority. By the end of this decade, we aim to have defeated disease as a horseman that can ravage societies unchecked.

Planet in Our Hands:

This decade isn't just about changing the physical landscape; it's about reshaping mindsets. By the end of twenty years, the world will be different. A new generation, born into a society where sustainability is not just a lesson but a way of life, will emerge. They will see themselves as stewards of the Earth, not its conquerors.

The Third Decade: Embracing Inclusion and Justice (Years 21-30)

As we move into the third decade, we confront an often overlooked but crucial element of transformation: social justice and inclusion. Without equity, sustainability remains out of reach. We must ensure that the benefits of this transformation are shared by all and that those who have been marginalized or displaced by climate change are fully integrated into this new world.

Key Actions:

- **Policies for Climate Refugees**: Establish international agreements that provide safe, legal pathways for climate refugees. Countries collaborate to build new communities that are sustainable, inclusive, and equipped with the resources people need to thrive. No one is left behind, and the global community comes together to support those forced from their homes.

- **Universal Basic Income (UBI) and Green Jobs Programs**: As automation and technology shift the job landscape, governments implement UBI programs to ensure that no one is left without means. Alongside this, green job initiatives train millions in renewable energy, sustainable construction, and ecological restoration. These programs focus on integrating displaced workers and creating new opportunities for those previously marginalized by the economy.

- **Global Misinformation Control Measures**: We tackle misinformation directly, creating international coalitions that regulate and monitor the spread of false information. Countries enact laws that require transparency from media and technology companies, ensuring that algorithms promote factual, unbiased content rather than divisive, misleading narratives.

By the end of this decade, the world will see not just environmental progress but social transformation. We'll have built a society where climate justice is central, and communities once at the margins will now stand at the heart of a global movement for a fairer, more inclusive planet.

The Fourth Decade: Mastering Technology (Years 31-40)

In the fourth decade, technology takes center stage as a tool for humanity's advancement. AI and biotechnology, once sources of fear and uncertainty, become allies in our quest for sustainability. We harness their power to solve problems that once seemed insurmountable, ensuring that AI does not become a force that undermines society but rather one that empowers it.

Key Actions:

- **Artificial Intelligence for Environmental Monitoring**: AI systems monitor ecosystems in real-time, providing data to predict and mitigate disasters like forest fires, floods, and hurricanes. These technologies help governments and communities respond proactively, protecting lives and resources.

- **Biotechnology for Sustainable Agriculture**: CRISPR technology and other gene-editing advancements are used to develop crops that are resilient to changing climates, require less water, and are naturally resistant to pests, reducing the need for harmful pesticides. Farmers gain access to these innovations, ensuring food security for millions.

- **Decentralized Clean Energy Grids**: Using blockchain and other emerging technologies, communities build decentralized energy grids that allow them to generate, share, and store renewable energy. This reduces reliance on centralized power sources and ensures that even remote areas have access to clean, affordable electricity.

By this point, technology is no longer seen as an abstract or distant concept but as an integrated part of daily life. It becomes the bridge between human aspirations and environmental harmony, allowing us to live more sustainably while enhancing our quality of life.

Planet in Our Hands:

The Fifth to Seventh Decades: Consolidation and Global Collaboration (Years 41-70)

These decades are about consolidating the gains we've made and building deep, global collaborations. With the world increasingly interconnected, nations work not in isolation but as a unified entity, sharing resources, knowledge, and technologies to solve global problems.

Key Actions:

- **Global Treaty for Ocean Conservation**: Nations come together to protect the world's oceans, agreeing on measures to prevent overfishing, reduce plastic waste, and conserve marine habitats. International task forces patrol waters, ensuring that these agreements are enforced and that the world's oceans remain healthy.

- **Intercontinental Networks of Green Cities**: Cities around the world establish green corridors, linking urban centers with nature reserves and rural communities. These cities share best practices in sustainable development, from transportation systems to energy management, creating a global network of cities committed to zero-emission living. These green cities become hubs of innovation and cultural exchange, where people come together to share solutions and build a sense of global unity. The cities of this era are designed with nature in mind, integrating green spaces, vertical gardens, and urban forests that purify the air and create habitats for wildlife.

- **Human Rights and Environmental Justice Coalition**: Governments, NGOs, and grassroots organizations form alliances to address injustices that intersect with environmental issues. These coalitions advocate for indigenous rights, land protection, and equitable resource distribution, ensuring that all voices are heard. The focus shifts towards protecting those communities most vulnerable to environmental harm, emphasizing the importance of justice and fairness in the face of global change.

As this phase concludes, humanity will have created a framework for international cooperation that goes beyond treaties and trade agreements; it will be a deep, systemic commitment to protect life on

Earth. These decades represent the point at which the horsemen of war and death are pushed to the margins, as nations unite not through military alliances but through a shared commitment to survival and prosperity.

The Final Decades: A Flourishing Planet (Years 71-100)

The final decades of the 100 Year Mission mark the era of flourishing—where humanity's relationship with the planet and each other reaches a harmonious balance. The foundations built in earlier decades allow societies to thrive sustainably, and the technologies mastered become tools of abundance and well-being rather than fear and control. It is a time where the four ancient horsemen—war, famine, disease, and death—along with their modern counterparts, climate change and AI, are not eliminated but are controlled, understood, and balanced.

Key Actions:

- **Rewilding Programs and Biodiversity Sanctuaries**: Vast areas of land and ocean are designated as protected sanctuaries, where ecosystems are allowed to regenerate without human interference. These rewilded spaces not only become safe havens for endangered species but also act as the lungs of the Earth, capturing carbon and restoring biodiversity on an unprecedented scale. Communities engage in these efforts, seeing the fruits of a century-long labor to heal the scars left by industrialization and deforestation.

- **Global Health and Well-being Initiatives**: Universal access to healthcare, clean water, and nutritious food becomes the norm. Communities are connected by networks that ensure every person's basic needs are met, regardless of location or economic status. Advances in medical technology and genetic research, made accessible to all, extend life spans and enhance quality of life, fulfilling the promise that some people alive today may indeed witness this new era unfold.

- **Cultural Renaissance and Global Collaboration**: With the basic needs of humanity met and environmental stability achieved, the focus shifts to cultural growth and creative expression. Societies celebrate their shared achievements through arts, music, and cultural festivals that cross borders and blend traditions. This global renaissance marks a time

Planet in Our Hands:

where humanity is not just surviving but truly flourishing—where innovation, collaboration, and celebration are everyday experiences.

By the end of the hundred year period, we will have overcome the horsemen that once threatened our existence. War, famine, disease, death, climate change, and AI no longer hold power over us; instead, they are understood, managed, and rendered harmless through our collective efforts. AI, once a fearsome tool of manipulation, becomes a partner in maintaining balance and peace, its algorithms designed to promote harmony and safeguard against the resurgence of the forces that once divided and threatened humanity.

A Note for Those Who Stand at the Beginning

As we stand at the threshold of this journey, it's important to recognize that this mission isn't about some far-off future disconnected from our lives. It begins now, with us. Those who are alive today, especially the young and those yet to be born in the coming years, may very well witness the culmination of this 100 Year Mission. The seeds we plant now will grow, and the generations that follow will tend and nurture them. This is a call to action and a promise—a promise that we have the power to build a world where harmony is possible, where the forces of destruction no longer hold sway, and where humanity can thrive in unity with the planet we call home – the only planet that is home to all of us. It's a reminder that the future is not a distant dream; it's an ongoing story that we are all a part of, one that starts with us and stretches beyond the horizon, linking our hands with those who will come long after we are gone.

I know that this may read like a utopian dream. Perhaps it is. But **climate change is not a dream, it is a nightmare, and it will destroy us if we do not start to act now.** If you do not believe me then look up the Thwaites Glacier. Look at the rise in sea levels if it slips into the ocean. Then look at what will happen when the ice that it is holding back slips into the ocean too. Where do you live? Do you live anywhere near the ocean? If so, you should know that before the end of this century your neighbourhood might no longer exist – unless something is done. **Sadly, the Thwaites Glacier is only one of the many climate nightmares facing us.**

Planet in Our Hands:

We don't like change. Especially if we feel that we're living in a safe and stable community. We tend to not want anything to disrupt that safe cocoon that we believe protects us from everything that is 'bad' out there. The reality is that, like it or not, change is coming. If we do nothing the world will get warmer and life will become difficult for all humans, regardless of where you live. If we make and control changes to our lifestyles, how we relate to each other and how we relate to the planet, well, that utopian dream may just become real.

This is where you come in.

The 100 Year Mission is our legacy. It's the roadmap to a future where life isn't just sustained but celebrated. It's a testament that we can, and will, defeat the horsemen—both ancient and new—by building a world defined not by fear but by hope, resilience, and the enduring strength of the human spirit.

The planet is in our hands.

If you wish to participate in this mission, then please join us at www.100yearmission.com.

If you think that this book is important then please tell your friends, relatives and associates about it. As a gift to all our grandchildren send them a copy or recommend that they buy their own copy.

Appendix:

Reflections on Faith, Science, and Humanity

The author of this book believes in God. This belief is not grounded in fact, nor is it grounded in blind faith but in a deep, personal exploration of the world through both the lens of science and spirituality. In a time when science and faith are often set in opposition, I find that they can, and should, complement each other. Many educated, science-based people might dismiss belief in God as unnecessary or unscientific. But I believe that faith in God helps fill the profound gaps in human understanding. It addresses those questions science has yet to answer, and perhaps never fully can.

Take the concept of multiple universes or branes, for example. These are ideas that exist largely within theoretical physics and cosmology, with limited empirical evidence to support them. Yet, many scientists accept the possibility of these things, despite the absence of direct proof. In the same way, belief in God fills the space where our understanding of the universe ends and the mysteries begin. It is no less fact-based than the belief in multiverses, extra dimensions or other speculative theories in cosmology.

At the heart of my belief is the idea that this world, this vast universe, is too finely tuned and too complex to be a mere accident. The idea that life emerged by chance, that the forces holding everything together have no guiding principle, seems as implausible as any belief in the divine. This conviction does not come from a rejection of science but an

Planet in Our Hands:

embrace of it. Science and faith, I believe, should be two sides of the same coin, each exploring different aspects of the human experience. So, I believe in God. Not because there is any logical reason to, but because I choose to.

Guided by the Teachings of Jesus

I am a Christian, and I accept the fact that my being a Christian is probably an accident of my birth. If I'd been born in Pakistan, I'd likely be Muslim, In India a Hindu, in Japan – Shinto. I've looked at the core beliefs and values of all the major religions but I'm not a theologian and I'm not about to start preaching. If you want to learn more about religion then consult with your local priest, Rabi or Imam. Or do some reading as I did. For me, the teachings of Jesus provide a moral and ethical framework that brings clarity to the chaos of the world. When Jesus taught us to pray, "Your kingdom come, your will be done, on earth as it is in heaven," he wasn't simply offering us a hopeful prayer for a distant future. He was calling us to bring the divine into our world here and now. The kingdom of God isn't something we wait for—it is something we strive to create through our actions, by living out the principles of love, justice, and compassion that Jesus taught.

Jesus' command to love our neighbors as ourselves is perhaps one of the most powerful guiding principles for humanity. If we took it seriously, we would treat everyone we meet—regardless of race, nationality, or social status—with kindness, respect and acceptance of our differences. This love is not limited to those within our own community or tribe. It's agape love, a love that shows universal empathy, kindness and goodwill for all mankind. Jesus challenged his followers to love everyone in the way we would love ourselves. This is not an easy task, but it is one that could transform the world if enough of us took it to heart.

Belief in God, especially a belief shaped by the teachings of Jesus, leads to a deep responsibility for the well-being of others and the world around us. It calls for an active faith, one that seeks to heal, to reconcile, and to create a better world for all people. This idea aligns with the idea of stewardship—our duty to care for the earth and for one another as we are all part of God's creation.

The Misuse of Religion

It would be dishonest not to acknowledge that religion has been misused throughout history. People have wrongly used religion to conquer, enslave, and oppress. From the Crusades to the transatlantic slave trade, religion has too often been invoked to justify actions that are directly opposed to the teachings of love, compassion, and justice. This, however, does not make religion itself bad. As C.S. Lewis wisely noted, "The abuse of a thing is no argument against its proper use."

Just because some have wielded religion as a tool for power, control, or violence does not mean that religion itself is inherently violent or oppressive. Human beings, in our flawed and selfish nature, are capable of distorting any good thing, and that includes faith. The actions of a few do not define the truth of a belief system. The core message of Christianity—love, forgiveness, and the pursuit of peace—remains unchanged, even if individuals or institutions have failed to live up to it.

C.S. Lewis also reflected on the importance of a moral framework provided by faith. In his book **Mere Christianity**, he said, "If you look for truth, you may find comfort in the end; if you look for comfort, you will not get either comfort or truth—only soft soap and wishful thinking to begin with and, in the end, despair." Religion, when followed sincerely and truthfully, provides a moral compass that guides us toward truth and away from the harmful distortions that power and greed can cause.

The misuse of religion by those who seek to enslave or conquer others stands in direct opposition to the teachings of Christ. Jesus did not call for domination or oppression—he called for love, humility, and service to others. His life and message were about lifting up the downtrodden, caring for the sick, and seeking justice for the oppressed. Therefore, true faith in God should lead us to reject violence and oppression, not embrace them.

The Intersection of Faith and Science

There are those within the Christian faith who struggle with the advances of science, particularly those that seem to contradict the Bible. Some believe that the world is only a few thousand years old, based on

Planet in Our Hands:

a literal reading of the book of Genesis. These beliefs conflict with overwhelming scientific evidence that the universe is billions of years old and that life evolved over long periods. It is important to acknowledge that science does not disprove the existence of God, but it does challenge certain traditional interpretations of scripture.

In my view, science is a tool—a way of understanding the physical world. The book of Genesis, and other religious texts, offer deeper truths about the human condition, morality, and our relationship with the divine. They are not meant to be scientific textbooks but spiritual guides. A belief in God and a belief in science are not mutually exclusive; they can work together to deepen our understanding of the world and our place in it.

The resistance to science within some religious communities is unfortunate because it creates unnecessary division. If we truly believe in God, then we should not fear science. If God created the universe, then studying that creation through science only brings us closer to understanding the divine. This does not mean we will ever have all the answers—far from it. But it does mean that we can appreciate the wonders of the universe while still holding fast to our faith.

The State of the World: Global Warming, War, and Misinformation

As we look at the state of the world today, it's easy to feel overwhelmed by the challenges we face. Global warming threatens the very future of humankind. Wars rage across the globe, fueled by greed, power, and misinformation. Technology, particularly AI, is increasingly used by those in power to deceive and manipulate for their own selfish gain. It is easy to become cynical or despairing in the face of such challenges.

However, a belief in God—and in the teachings of Christ—offers hope. The call to love one another, to love our neighbors as ourselves, is more important now than ever. If we can find a way to truly live by this principle, we can begin to address the root causes of the problems we face. War, environmental destruction, and the abuse of technology all stem from a lack of love, a lack of concern for the well-being of others.

We are called to act—not to wait passively for someone else to solve these problems but to work for the survival and flourishing of all humanity. This means advocating for policies that protect the environment, seeking peaceful resolutions to conflicts, and ensuring that technology is used ethically for the common good. Our faith can inspire us to take these actions, guided by the belief that we are all part of a greater whole.

Carl Sagan's famous description of Earth as a "pale blue dot" serves as a humbling reminder of our place in the universe. From a distance, our planet is just a small, fragile speck in the vastness of space. Yet, it is home to everything we know and love. If we truly understood how precious and rare life is, perhaps we would treat the earth and each other with more care. Faith teaches us to see the sacredness of life, and science reveals the delicate balance that sustains it. Together, they urge us to act with wisdom and compassion.

Conclusion: Love as the Solution

At the core of my faith is the belief that love is the solution to the collective challenges facing mankind. Whether you are Christian, Muslim, Jewish, Hindu, or atheist, the principle of love for one another is a universal truth. It transcends religious boundaries and speaks to the very heart of what it means to be human. Jesus's teachings on love are not just for Christians—they are for all of humanity.

If we can find a way to love one another, to see the humanity in each person we encounter, we can begin to solve the problems that threaten our world. Love is not a passive emotion; it is an active force that compels us to work for justice, for peace, and for the survival of our planet.

I believe that humanity can be "saved" from the challenges we face if we choose to love one another. This doesn't mean that everything will be easy or that the road ahead will be without hardship. But it does mean that there is hope, and that each of us has a role to play in creating a better future.

In the end, faith in God and in the teachings of Christ is not about escaping from the world—it is about transforming it. It is about

Planet in Our Hands:

bringing the kingdom of God to earth, as it is in heaven. And at the heart of this transformation is love—the kind of love that Jesus taught, the kind that sees no boundaries, no divisions, only the shared humanity of all people.

One of my favorite passages of scripture is found in 1 Corinthians chapter 13. It ends with three words faith, hope and love. I have used these words for many years as my personal motto. Let us choose to live by this love, for in doing so, we will change the world.

About Max Wynter

Max is about to complete his 7th decade and, in those years, he has lived a number of lives. As an undergraduate at the Mona Campus of the University of the West Indies Max's interest in Physics led to him receiving the department prize in 1974. He followed this with a postgraduate diploma in Applied Physics. This diploma was a step to a master's degree in applied physics. His master's degree project was to computerize the University's laser radar research station, at the time, the largest such facility in the world. The Physics department used this facility to research aerosols, fine solid particles in the Earth's atmosphere. This is where Max's interest in air pollution began. Then work and life got in the way. Max did not complete that degree.

Instead, he focused on his work as an Electronic Engineer at the Gleaner Company, one of the world's oldest newspaper publishers. This is where his interest in honest, fact-based news was sparked, and he became a self-described "news junkie". Eventually he became the Managing Director and CEO of the Jamaica Observer, another national daily newspaper and competitor to the Gleaner Company.

In between the Gleaner and the Observer Max partnered with a friend to start a computer company and then became the President of the Jamaica Computer Society. He left the computer company to become the CEO of Jamaica Digiport International – a telephone company. At the Jamaica Computer Society Max founded the Jamaica Computer Society Education Foundation with a mission of having a computerized classroom in every high school in Jamaica by the year 2000. This goal was achieved, and these classrooms were used to teach students computer science as well as math and the general sciences. At Jamaica Digiport, Max started a telemarketing call centre with 15 workstations, a pilot project to show that this industry was viable in Jamaica. The industry now employs over 40,000 persons in Jamaica.

While at Jamaica Digiport he was invited to be one of the first Commissioners of The Global Information Infrastructure Commission and sat on the commission for 4 years. In the 1990's he traveled the world speaking on the use of information technology in national development and in education.

He does not consider himself to be a physicist, electronic engineer, computer engineer, telecommunications executive, newspaper publisher or author although, in his career, he has been all of those. Instead, he reads a lot and, from what he has read, he has grave concerns about the future of mankind. This book is his attempt to communicate important messages in one easy to read book.

Max completed his education with an MBA and has strong views on business ethics, corporate social responsibility, our common humanity and the only planet that we all call home. He is a businessman and lives with his wife Anne in Canada.

Bibliography

1. Collins, Francis. "The Language of God: A Scientist Presents Evidence for Belief". Free Press, 2006.

2. Diamond, Jared. "Guns, Germs, and Steel: The Fates of Human Societies". W.W. Norton & Company, 1997.

3. Doudna, Jennifer A., and Samuel H. Sternberg. "A Crack in Creation: Gene Editing and the Unthinkable Power to Control Evolution". Houghton Mifflin Harcourt, 2017.

4. Gates, Bill. "How to Avoid a Climate Disaster: The Solutions We Have and the Breakthroughs We Need". Knopf, 2021.

5. Goodell, Jeff. "The Heat Will Kill You First: Life and Death on a Scorched Planet". Little, Brown and Company, 2023.

6. Harari, Yuval Noah. "Homo Deus: A Brief History of Tomorrow". Harper Perennial, 2017.

7. Harari, Yuval Noah. "Sapiens: A Brief History of Humankind". Harper Perennial, 2015.

8. Harari, Yuval Noah. "Nexus". McClelland & Stewart, 2024.

9. Harari, Yuval Noah. "21 Lessons for the 21st Century". Spiegel & Grau, 2018.

10. Hawking, Stephen. "A Brief History of Time". Bantam Books, 1988.

11. Hawking, Stephen. "The Theory of Everything: The Origin and Fate of the Universe". Phoenix Books, 2002.

12. Hayhoe, Katharine. "Saving Us: A Climate Scientist's Case for Hope and Healing in a Divided World". Atria/One Signal Publishers, 2021.

13. Holloway, Richard. "A Little History of Religion". Yale University Press, 2016.

Planet in Our Hands:

14. Kolbert, Elizabeth. "The Sixth Extinction: An Unnatural History". Henry Holt and Co., 2014.
15. Randall, Lisa. "Warped Passages: Unraveling the Mysteries of the Universe's Hidden Dimensions". Ecco, 2005.
16. Robinson, Kim Stanley. "2312". Orbit, 2012.
17. Robinson, Kim Stanley. "The Ministry for the Future". Orbit, 2020.
18. Sinclair, David A., and Matthew LaPlante. "Lifespan: Why We Age—and Why We Don't Have To". Atria Books, 2019.
19. Suleyman, Mustafa. "The Coming Wave: Technology, Power, and the Twenty-first Century's Greatest Dilemma". Crown Publishing, 2023.
20. Tyson, Neil deGrasse. "Starry Messenger: Cosmic Perspectives on Civilization". Henry Holt and Co., 2022.
21. Lewis, C. S. "Mere Christianity". HarperOne, 1952.

Additional Resources

Please note that any of the Web addresses or links below may have changed since publication and may no longer be valid.

22. ProPublica - Independent investigative journalism. -
https://www.propublica.org.
23. Associated Press - International and independent news agency. -
https://www.ap.org.
24. BBC News - Global news coverage. -
https://www.bbc.com/news.
25. Snopes - Fact-checking resource for verifying viral claims. -
https://www.snopes.com.
26. PolitiFact - Fact-checking organization. -
https://www.politifact.com.
27. Nature Journal - Peer-reviewed scientific journal. -
https://www.nature.com.
28. The Lancet - Peer-reviewed medical journal. -
https://www.thelancet.com.

29. Al Jazeera - International news coverage. - https://www.aljazeera.com.
30. Reuters - Global news organization. - https://www.reuters.com.
31. The Guardian - International news coverage. - https://www.theguardian.com.
32. Deepfake Technology Articles - Various sources discussing the implications of deepfake technology and AI-generated misinformation.